Test Yourself

College Algebra

Joan Van Glabek, Ph.D.
Department of Mathematics
Edison Community College
Naples, FL

Contributing Editors

David Ludwig, Ph.D.
Oakland, NJ

John Dauns, Ph.D.
Department of Mathematics
Tulane University
New Orleans, LA

Tony Julianelle, Ph.D.
Department of Mathematics
University of Vermont
Burlington, VT

NTC LearningWorks
NTC/Contemporary Publishing Group

Library of Congress Cataloging-in-Publication Data

Van Glabek, Joan.
    College algebra / Joan Van Glabek ; contributing editors, David
Ludwig, John Dauns, Tony Julianelle.
        p.   cm. — (Test yourself)
    ISBN 0-8442-2385-9
    1. Algebra—Examinations, questions, etc.   I. Ludwig, David.
II. Dauns, John.   III. Julianelle, Tony.   IV. Title.   V. Series:
Test yourself (Lincolnwood, Ill.)
QA157.V35   1998
512.9—dc21
                                                98-13196
                                                CIP

A *Test Yourself Books, Inc.* Project

Published by NTC LearningWorks
A division of NTC/Contemporary Publishing Group, Inc.
4255 West Touhy Avenue, Lincolnwood (Chicago), Illinois 60646-1975 U.S.A.
Printed in the United States of America
International Standard Book Number: 0-8442-2385-9
 18 17 16 15 14 13 12 11 10 9 8 7 6 5 4 3 2 1

Contents

Preface

These test questions and answers have been written to help you more fully understand the material in a college algebra course. Working through these test questions prior to a test will help you pinpoint areas that need further study. Each answer is followed by a parenthetical topic description that corresponds to a section in your college algebra textbook. Return to your textbook as needed to reread and review areas where you feel unsure about your work. See your professor for additional help, and/or get together with a study group to complete your preparation for each test. Work some problems from the test book every day. This will allow you time to properly prepare for your tests.

Practice and a positive attitude (just keep telling yourself how well you are doing) will help you succeed. Good luck with your study of algebra.

Joan Van Glabek, Ph.D.
Naples, FL

How to Use This Book

This "Test Yourself" book is part of a unique series designed to help you improve your test scores on almost any type of examination you will face. Too often, you will study for a test—quiz, midterm, or final—and come away with a score that is lower than anticipated. Why? Because there is no way for you to really know how much you understand a topic until you've taken a test. The *purpose* of the test, after all, is to test your complete understanding of the material.

The "Test Yourself" series offers you a way to improve your scores and to actually test your knowledge at the time you use this book. Consider each chapter a diagnostic pretest in a specific topic. Answer the questions, check your answers, and then give yourself a grade. Then, and only then, will you know where your strengths and, more important, weaknesses are. Once these areas are identified, you can strategically focus your study on those topics that need additional work.

Each book in this series presents a specific subject in an organized manner, and although each "Test Yourself" chapter may not correspond to exactly the same chapter in your textbook, you should have little difficulty in locating the specific topic you are studying. Written by educators in the field, each book is designed to correspond, as much as possible, to the leading textbooks. This means that you can feel confident in using this book and that regardless of your textbook, professor, or school, you will be much better prepared for anything you will encounter on your test.

Each chapter has four parts:

 Brief Yourself. All chapters contain a brief overview of the topic that is intended to give you a more thorough understanding of the material with which you need to be familiar. Sometimes this information is presented at the beginning of the chapter, and sometimes it flows throughout the chapter, to review your understanding of various *units* within the chapter.

 Test Yourself. Each chapter covers a specific topic corresponding to one that you will find in your textbook. Answer the questions, either on a separate page or directly in the book, if there is room.

 Check Yourself. Check your answers. Every question is fully answered and explained. These answers will be the key to your increased understanding. If you answered the question incorrectly, read the explanations to *learn* and *understand* the material. You will note that at the end of every answer you will be referred to a specific subtopic within that chapter, so you can focus your studying and prepare more efficiently.

 Grade Yourself. At the end of each chapter is a self-diagnostic key. By indicating on this form the numbers of those questions you answered incorrectly, you will have a clear picture of your weak areas.

There are no secrets to test success. Only good preparation can guarantee higher grades. By utilizing this "Test Yourself" book, you will have a better chance of improving your scores and understanding the subject more fully.

Basic Algebra Review

1

 ## Brief Yourself

The topics in this chapter comprise a review of topics from intermediate algebra. These topics include real numbers, integer exponents, polynomials, factoring, rational expressions, radicals and rational exponents, and complex numbers.

The real numbers are generally described using set notation. You should be familiar with the following symbols:

$A \cup B$	Union. A union B is the set that contains all elements in set A *or* set B.
$A \cap B$	Intersection. A intersection B is the set that contains all elements found in both set A *and* set B.
$A \subseteq B$	Subset. A is a subset of B if every element of set A is an element of set B.
$A \nsubseteq B$	Not a subset. A is not a subset of B if some element of A is not an element of set B.
$a \in A$	Element of. a is a member of set A.
$a \notin B$	Not an element of. a is not an element of B.

The usual operations with real numbers (addition, subtraction, multiplication, division) can be combined in one problem; in such a case, the rules for order of operations dictate how to proceed (parentheses, exponents, multiplication and division, addition and subtraction).

The following properties of real numbers will be used throughout *College Algebra*:

Commutative Property of Addition	$a + b = b + a$
Commutative Property of Multiplication	$ab = ba$
Associative Property of Addition	$a + (b + c) = (a + b) + c$
Associative Property of Multiplication	$a(bc) = (ab)c$
Additive Identity Property	$a + 0 = a = 0 + a$
Multiplicative Identity Property	$a(1) = a = 1(a)$
Additive Inverse Property	$a + (-a) = 0$
Multiplicative Inverse Property	$a\left(\dfrac{1}{a}\right) = 1 \quad \text{for } a \neq 0$
Distributive Property	$a(b + c) = ab + ac$

The following list summarizes the properties of exponents and the definitions associated with integer exponents m and n:

$$x^m x^n = x^{m+n}$$

$$\left(x^m\right)^n = x^{mn}$$

$$(xy)^n = x^n y^n$$

$$\left(\frac{x}{y}\right)^n = \frac{x^n}{y^n}, \; y \neq 0$$

$$\frac{x^m}{x^n} = x^{m-n}, \; x \neq 0$$

$$x^0 = 1, \; x \neq 0$$

$$x^{-n} = \frac{1}{x^n}$$

This chapter also contains problems that review operations with polynomials. Some products occur frequently in algebra; their formulas should be memorized.

$$(x \pm y)^2 = x^2 \pm 2xy + y^2$$

$$(x + y)^3 = x^3 + 3x^2 y + 3xy^2 + y^3$$

$$(x - y)^3 = x^3 - 3x^2 y + 3xy^2 - y^3$$

$$(x + y)(x - y) = x^2 - y^2$$

To factor a polynomial completely means to write it as a product of prime factors. Always begin by factoring out the greatest common factor. Factoring techniques include trial and error and factoring by grouping. The formulas shown below should be memorized:

Binomials

$$x^2 - y^2 = (x - y)(x + y)$$

$$x^3 - y^3 = (x - y)(x^2 + xy + y^2)$$

$$x^3 + y^3 = (x + y)(x^2 - xy + y^2)$$

Trinomials

$$x^2 + 2xy + y^2 = (x + y)^2$$

$$x^2 - 2xy + y^2 = (x - y)^2$$

Rational expressions (fractions) can be combined by the usual operations of addition, subtraction, multiplication, and division. A rational expression is undefined for all values that make the denominator equal 0. To add or subtract rational expressions, first find the least common denominator. Before multiplying rational expressions, factor each numerator and denominator and look for common factors to cancel.

When working with radicals, be sure to simplify answers—that is, powers in the radicand must be less than the index of the radical, and no fractions can appear under the radical. The definition of a fractional exponent should be memorized:

$$x^{p/q} = \sqrt[q]{x^p} = \left(\sqrt[q]{x}\right)^p$$

Complex numbers are numbers that can be written as $a + bi$, where $i = \sqrt{-1}$. Since $i = \sqrt{-1}$, $i^2 = -1$. Complex numbers can be combined by the operations of addition, subtraction, multiplication, and division and answers should be written in simplified form, such that the highest power of i is 1 and i does not appear in the denominator. If the denominator of a fraction is a monomial containing i, multiply by $\frac{i}{i}$ and simplify. If the denominator is a binomial of the form $a + bi$ multiply by $\frac{a - bi}{a - bi}$ and simplify.

Test Yourself

1. Which numbers in the set
$$\{12, -1, \frac{12}{7}, \frac{2}{3}, \sqrt{3}, 0, 5\} \text{ are:}$$

   (a) natural numbers?
   (b) integers?
   (c) rational numbers?
   (d) irrational numbers?

2. Which numbers in the set
$$\{0.35, -\pi, \frac{4}{2}, 3, \sqrt{6}, 0\} \text{ are:}$$

   (a) whole numbers?
   (b) rational numbers?
   (c) irrational numbers?

Perform the indicated operations.

3. $|-2| - |-3|$

4. $-\frac{3}{4} \div \frac{6}{28}$

5. $4[2 - 3(5 - 12)]$

6. $\dfrac{-2^4 + (-2)^2}{-4 - (2 - 2)}$

7. $-\frac{2}{3} - \left(\frac{1}{6} - \frac{2}{9}\right)$

Identify the property or properties of real numbers illustrated by each equation.

8. $(2x)(y) = 2(xy)$

9. $\dfrac{3}{x + 2} \cdot \dfrac{x + 2}{3} = 1, \; x \neq -2$

10. $(x + 3)(x - 1) = (x - 1)(x + 3)$

11. $(4x + 6) + 2x = (4x + 2x) + 6$

12. $4(x + y) = 4x + 4y$

Simplify the expression.

13. $\dfrac{(3x)^{-3}}{(2y)^{-2}}$

14. $\dfrac{3^{-3} x^5 y^{-3}}{9^{-2} x^4 y^2}$

15. $\left(\dfrac{3x^3 y^2}{2x^{-2} y^5}\right)^{-4}$

16. Write 2,780,000 in scientific notation.

17. Write 6.05×10^{-3} in decimal form.

Perform the indicated operations.

18. $(3x^2 - 4x + 12) - (4x^3 - 12x + 11)$

19. $5y - [3y - (8y + 4)]$

20. $(4x-3)(2x+5)$

21. $(3a+2b)^2$

22. $(4-5x)^2$

23. $(2m-1)^3$

24. $(3+x)^3$

25. $(2x+3y)(2x-3y)$

26. $[(x-2)+y][(x-2)-y]$

27. $\dfrac{6x^4-9x^3-12x+3}{3x^4}$

28. $\dfrac{x^2-4}{x+1}$

29. $\dfrac{2x^3-4x^2-15x+5}{x^2-x-8}$

Factor each polynomial completely.

30. $4x^2-81y^4$

31. $27x^3+64y^3$

32. $9x^2-6x+1$

33. $(p-1)^2-36$

34. $8x^3-20x^2-48x$

35. $2x^4-8x^3-8x^2+32x$

36. $3x^3+x^2+12x+4$

37. $5x(3x-1)+(3x-1)^2$

38. Determine the values for which $\dfrac{x-2}{x^2-4}$ is undefined.

Perform the indicated operations and simplify.

39. $\dfrac{4x^2-4x+1}{4x^2-1}\cdot\dfrac{2x+1}{x-2}$

40. $\dfrac{x^2+xy-2y^2}{x^3+x^2y}\div\dfrac{x^2+3xy+2y^2}{x}$

41. $\dfrac{2}{x^2-x-6}-\dfrac{x}{x^2-5x+6}$

42. $\dfrac{-21}{x^2+3x-10}+\dfrac{3}{x-2}$

43. Simplify $\dfrac{\dfrac{1}{x}-\dfrac{1}{2}}{\dfrac{2}{x}+1}$.

44. Simplify $\dfrac{\dfrac{1}{x+h+1}-\dfrac{1}{x+1}}{h}$.

45. Simplify $\dfrac{\dfrac{1}{x-5}-\dfrac{1}{x+4}}{\dfrac{2x^2-8}{x^2-x-20}}$.

Simplify each expression.

46. $-\sqrt{54x^3}$

47. $\sqrt[4]{32x^9y^4}$

48. $\sqrt[3]{\dfrac{12x^5y^4}{16xy^2}}$

49. $32^{3/5}$

50. $25^{-1/2}$

Perform the indicated operations and simplify.

51. $x^{1/2}\cdot x^{2/3}$

52. $\dfrac{x^{2/3}y^{1/2}}{(xy)^{1/3}}$

53. Rationalize the denominator: $\dfrac{3}{\sqrt{8}}$.

54. Rationalize the numerator: $\dfrac{\sqrt{5}-3}{4}$.

Perform the indicated operations and simplify.

55. $2\sqrt{50}-5\sqrt{8}$

56. $4\sqrt[3]{16}+2\sqrt[3]{54}$

57. $3\sqrt{2}(5-2\sqrt{2})$

58. $(\sqrt{5}+2\sqrt{3})(3\sqrt{5}-4\sqrt{3})$

Perform the indicated operations and simplify.

59. $(8 - 5i) - (3 - 4i)$

60. $(2 + 3i)(2 - 3i)$

61. $\sqrt{-12} + 5\sqrt{-27}$

62. $\dfrac{5 - 8i}{2i}$

63. $\dfrac{6}{4 - i}$

 ## Check Yourself

1. (a) 5, 12

   (b) 5, 12, -1, 0

   Note that every natural number is an integer.

   (c) 5, 12, -1, 0, $\dfrac{12}{7}$, $\dfrac{2}{3}$

   Note that every integer is a rational number.

   (d) $\sqrt{3}$

   (The real numbers)

2. (a) $\dfrac{4}{2}$, 3, 0

   Note that 0 is a whole number, but not a natural number. Since 4/2 = 2, it is included in the set of whole numbers.

   (b) 0, 0.35, $\dfrac{4}{2}$, 3, 0

   Decimals can be written as fractions ($0.35 = \dfrac{35}{100}$), so are included in the set of rational numbers.

   (c) $-\pi$, $\sqrt{6}$

   (The real numbers)

3. $|-2| - |-3| = 2 - 3 = -1$

   (The real numbers)

4. $-\dfrac{3}{4} \div \dfrac{6}{28} = -\dfrac{3}{4} \cdot \dfrac{28}{6}$

   Invert and multiply.

   $= -\dfrac{\overset{1}{\cancel{3}}}{\underset{1}{\cancel{4}}} \cdot \dfrac{\overset{7}{\cancel{28}}}{\underset{2}{\cancel{6}}}$

   Cancel.

   $= -\dfrac{7}{2}$

   (The real numbers)

5. $4[2 - 3(5 - 12)] = 4[2 - 3(-7)]$

   Work inside parentheses.

   $= 4[2 + 21]$

   Multiply.

$$= 4[23] \qquad \text{Work inside brackets.}$$

$$= 92 \qquad \text{Multiply.}$$

(The real numbers)

6. $\dfrac{-2^4 + (-2)^2}{-4 - (2-2)} = \dfrac{-16+4}{-4-(0)}$ Note that $-2^4 = -1 \cdot 2^4$ while $(-2)^2 = -2 \cdot -2$.

$$= \dfrac{-12}{-4} = 3$$

(The real numbers)

7. $-\dfrac{2}{3} - \left(\dfrac{1}{6} - \dfrac{2}{9}\right)$ Use 18 as the LCD (least common denominator).

$$= -\dfrac{12}{18} - \left(\dfrac{3}{18} - \dfrac{4}{18}\right) \qquad \text{Write each fraction using 18 as the denominator.}$$

$$= -\dfrac{12}{18} - \left(-\dfrac{1}{18}\right) \qquad \text{Work inside parentheses first.}$$

$$= -\dfrac{12}{18} + \dfrac{1}{18} = -\dfrac{11}{18}$$

(The real numbers)

8. Associative property of multiplication

(The real numbers)

9. Inverse property of multiplication

(The real numbers)

10. Commutative property of multiplication

(The real numbers)

11. $(4x + 6) + 2x = 4x + (6 + 2x)$ Associative property of addition.

$$= 4x + (2x + 6) \qquad \text{Commutative property of addition.}$$

$$= (4x + 2x) + 6 \qquad \text{Associative property of addition.}$$

(The real numbers)

12. Distributive property

(The real numbers)

13. $\dfrac{(3x)^{-3}}{(2y)^{-2}} = \dfrac{(2y)^2}{(3x)^3}$ Use $x^{-n} = \dfrac{1}{x^n}$

$\qquad = \dfrac{2^2 y^2}{3^3 x^3}$ Use $(xy)^n = x^n y^n$

$\qquad = \dfrac{4y^2}{27x^3}$

(Integer exponents and scientific notation)

14. $\dfrac{3^{-3} x^5 y^{-3}}{9^{-2} x^4 y^2} = \dfrac{9^2 x^5}{3^3 x^4 y^2 y^3}$ Use $x^{-n} = \dfrac{1}{x^n}$

$\qquad = \dfrac{81x}{27y^5}$ $\dfrac{x^5}{x^4} = x^{5-4} = x^1 \qquad y^2 \cdot y^3 = y^{2+3} = y^5$

$\qquad = \dfrac{3x}{y^5}$

(Integer exponents and scientific notation)

15. $\left(\dfrac{3x^3 y^2}{2x^{-2} y^5}\right)^{-4} = \left(\dfrac{3x^5}{2y^3}\right)^{-4}$ Simplify inside the parentheses.

$\qquad = \left(\dfrac{2y^3}{3x^5}\right)^4$ Use $\left(\dfrac{x}{y}\right)^{-n} = \left(\dfrac{y}{x}\right)^n$.

$\qquad = \dfrac{2^4 y^{12}}{3^4 x^{20}}$ Use $\left(\dfrac{x}{y}\right)^n = \dfrac{x^n}{y^n}$.

$\qquad = \dfrac{16y^{12}}{81x^{20}}$

(Integer exponents and scientific notation)

16. $2{,}780{,}000 = 2.78 \times 10^6$ Note that the decimal moved 6 places to the left so the exponent on 10 is 6.

(Integer exponents and scientific notation)

17. $6.05 \times 10^{-3} = 0.00605$ The exponent of -3 means that the decimal point should be moved 3 places to the left.

(Integer exponents and scientific notation)

18. $(3x^2 - 4x + 12) - (4x^3 - 12x + 11)$

$= 3x^2 - 4x + 12 - 4x^3 + 12x - 11$ Change the sign of each term in the polynomial being subtracted.

$= -4x^3 + 3x^2 + 8x + 1$ Combine like terms.

(Polynomials)

19. $5y - [3y - (8y + 4)]$

$= 5y - [3y - 8y - 4]$

$= 5y - [-5y - 4]$ Combine like terms.

$= 5y + 5y + 4$ Change signs to subtract.

$= 10y + 4$ Combine like terms.

(Polynomials)

20. $(4x - 3)(2x + 5) = 8x^2 + 20x - 6x - 15$ Use the distributive property or FOIL.

$= 8x^2 + 14x - 15$

(Polynomials)

21. Use the formula $(x + y)^2 = x^2 + 2xy + y^2$:

$(3a + 2b)^2 = (3a)^2 + 2(3a)(2b) + (2b)^2 = 9a^2 + 12ab + 4b^2$

(Polynomials)

22. Use the formula $(x - y)^2 = x^2 - 2xy + y^2$:

$(4 - 5x)^2 = (4)^2 - 2(4)(5x) + (5x)^2 = 16 - 40x + 25x^2$

(Polynomials)

23. Use the formula $(x - y)^3 = x^3 - 3x^2y + 3xy^2 - y^3$:

$(2m - 1)^3 = (2m)^3 - 3(2m)^2(1) + 3(2m)(1)^2 - (1)^3$

$= 8m^3 - 12m^2 + 6m - 1$

(Polynomials)

24. $(3 + x)^3 = (3)^3 + 3(3)^2x + 3(3)x^2 + (x)^3$

$= 27 + 27x + 9x^2 + x^3$

(Polynomials)

25. Use the formula $(x + y)(x - y) = x^2 - y^2$:

$$(2x + 3y)(2x - 3y) = (2x)^2 - (3y)^2$$

$$= 4x^2 - 9y^2$$

(Polynomials)

26. $[(x - 2) + y][(x - 2) - y]$

$$= (x - 2)^2 - (y)^2 \qquad \text{Use the formula } (x + y)(x - y) = x^2 - y^2.$$

$$= x^2 - 4x + 4 - y^2 \qquad \text{Use the formula } (x - y)^2 = x^2 - 2xy + y^2.$$

(Polynomials)

27. $\dfrac{6x^4 - 9x^3 - 12x + 3}{3x^4} = \dfrac{6x^4}{3x^4} - \dfrac{9x^3}{3x^4} - \dfrac{12x}{3x^4} + \dfrac{3}{3x^4}$ Divide each term of the numerator by the denominator.

$$= 2 - \frac{3}{x} - \frac{4}{x^3} + \frac{1}{x^4} \qquad \text{Simplify each term.}$$

(Polynomials)

28.

$$
\begin{array}{r}
x - 1 \\
x + 1 \overline{) x^2 + 0x - 4} \\
\underline{x^2 + \ x} \\
-x - 4 \\
\underline{-x - 1} \\
-3
\end{array}
$$

The answer may be written as $x - 1 - \dfrac{3}{x + 1}$ or $x - 1 \ \text{R}(-3)$.

(Polynomials)

29.

$$
\begin{array}{r}
2x - 2 \\
x^2 - x - 8 \overline{) 2x^3 - 4x^2 - 15x + 5} \\
\underline{2x^3 - 2x^2 - 16x} \\
-2x^2 + \ x + 5 \\
\underline{-2x^2 + \ 2x + 16} \\
-x - 11
\end{array}
$$

$2x - 2 + \dfrac{-x - 11}{x^2 - x - 8}$ or $2x - 2 - \dfrac{x + 11}{x^2 - x - 8}$ or $2x - 2 R(-x - 11)$.

(Polynomials)

30. $4x^2 - 81y^4$ Use $x^2 - y^2 = (x - y)(x + y)$.

    $= (2x)^2 - (9y^2)^2$

    $= (2x - 9y^2)(2x + 9y^2)$

    (Factoring)

31. $27x^3 + 64y^3$ Use $x^3 + y^3 = (x + y)(x^2 - xy + y^2)$.

    $= (3x)^3 + (4y)^3$

    $= (3x + 4y)(9x^2 - 12xy + 16y^2)$

    (Factoring)

32. $9x^2 - 6x + 1$ Use $x^2 - 2xy + y^2 = (x - y)^2$.

    $= (3x - 1)^2$

    (Factoring)

33. $(p - 1)^2 - 36$ Use $x^2 - y^2 = (x - y)(x + y)$.

    $= (p - 1)^2 - (6)^2$

    $= [(p - 1) - 6][(p - 1) + 6]$

    $= (p - 7)(p + 5)$

    (Factoring)

34. $8x^3 - 20x^2 - 48x$

    $= 4x(2x^2 - 5x - 12)$ Factor out $4x$.

    $= 4x(2x + 3)(x - 4)$ Use trial and error to factor the trinomial.

    (Factoring)

35. $2x^4 - 8x^3 - 8x^2 + 32x$

    $= 2x(x^3 - 4x^2 - 4x + 16)$ Factor out $2x$.

    $= 2x[x^2(x - 4) - 4(x - 4)]$ Factor by grouping.

    $= 2x[(x^2 - 4)(x - 4)]$ Common factor.

    $= 2x(x - 2)(x + 2)(x - 4)$ $x^2 - 4 = (x)^2 - (2)^2 = (x - 2)(x + 2)$.

    (Factoring)

36. $3x^3 + x^2 + 12x + 4$

     $= x^2(3x+1) + 4(3x+1)$ Factor by grouping.

     $= (3x+1)(x^2+4)$

Note that x^2+4 (a sum of two squares) cannot be factored further.

   (Factoring)

37. $5x(3x-1) + (3x-1)^2$

     $= (3x-1)(5x+3x-1)$ Factor out $(3x-1)$.

     $= (3x-1)(8x-1)$ Add like terms.

   (Factoring)

38.   When $x^2 - 4 = 0$, the fraction is undefined.

       $x^2 - 4 = 0$

   $(x-2)(x+2) = 0$ Factor.

   $x - 2 = 0$ or $x + 2 = 0$ Set each factor equal to 0.

     $x = 2$ or $x = -2$

The rational expression is undefined when $x = 2$ or -2.

   (Rational expressions)

39. $\dfrac{4x^2 - 4x + 1}{4x^2 - 1} \cdot \dfrac{2x+1}{x-2}$

     $= \dfrac{(2x-1)^2}{(2x-1)(2x+1)} \cdot \dfrac{2x+1}{x-2}$ Factor.

     $= \dfrac{2x-1}{x-2}$ Divide out common factors.

   (Rational expressions)

40. $\dfrac{x^2 + xy - 2y^2}{x^3 + x^2 y} \div \dfrac{x^2 + 3xy + 2y^2}{x}$

     $= \dfrac{(x+2y)(x-y)}{x^2(x+y)} \cdot \dfrac{x}{(x+2y)(x+y)}$ Invert divisor and multiply. Factor.

$$= \frac{(x-y)}{x(x+y)(x+y)} \quad \text{or} \quad \frac{x-y}{x(x+y)^2}$$

(Rational expressions)

41. $\dfrac{2}{x^2-x-6} - \dfrac{x}{x^2-5x+6}$

$= \dfrac{2}{(x-3)(x+2)} - \dfrac{x}{(x-3)(x-2)}$ Factor each denominator.

$= \dfrac{2}{(x-3)(x+2)} \cdot \dfrac{x-2}{x-2} - \dfrac{x}{(x-3)(x-2)} \cdot \dfrac{x+2}{x+2}$ The LCD is $(x-3)(x+2)(x-2)$.

$= \dfrac{2(x-2)-x(x+2)}{(x-3)(x+2)(x-2)}$

$= \dfrac{2x-4-x^2-2x}{(x-3)(x+2)(x-2)}$ Multiply and combine the numerators.

$= \dfrac{-x^2-4}{(x-3)(x+2)(x-2)} \quad \text{or} \quad - \dfrac{x^2+4}{(x-3)(x+2)(x-2)}$

(Rational expressions)

42. $\dfrac{-21}{x^2+3x-10} + \dfrac{3}{x-2}$

$= \dfrac{-21}{(x+5)(x-2)} + \dfrac{3}{x-2} \cdot \dfrac{x+5}{x+5}$ The LCD is $(x+5)(x-2)$.

$= \dfrac{-21+3x+15}{(x+5)(x-2)}$ Combine the numerators.

$= \dfrac{3x-6}{(x+5)(x-2)}$ Combine like terms.

$= \dfrac{3(x-2)}{(x+5)(x-2)}$ Factor the numerator.

$= \dfrac{3}{x+5}$ Divide out common factors.

(Rational expressions)

43. $\dfrac{\dfrac{1}{x}-\dfrac{1}{2}}{\dfrac{2}{x}+\dfrac{1}{1}}$ Note the LCD is $2x$.

$$= \frac{2x \cdot \dfrac{1}{x} - \dfrac{1}{2} \cdot 2x}{2x \cdot \dfrac{2}{x} + \dfrac{1}{1} \cdot 2x}$$

Multiply each term by the LCD.

$$= \frac{2 - x}{4 + 2x}$$

Divide out common factors.

(Rational expressions)

44. $\dfrac{\dfrac{1}{x+h+1} - \dfrac{1}{x+1}}{h}$

Use $(x+h+1)(x+1)$ as the LCD for the numerator.

$$= \frac{\dfrac{x+1}{x+1} \cdot \dfrac{1}{x+h+1} - \dfrac{1}{x+1} \cdot \dfrac{x+h+1}{x+h+1}}{\dfrac{h}{1}}$$

$$= \frac{x+1-x-h-1}{(x+1)(x+h+1)} \cdot \frac{1}{h}$$

Invert and multiply since dividing by $h/1$ is equivalent to multiplying by $1/h$.

$$= \frac{-h}{(x+1)(x+h+1)} \cdot \frac{1}{h}$$

$x+1-x-h-1 = -h$.

$$= \frac{-1}{(x+1)(x+h+1)}$$

Divide out common factors.

(Rational expressions)

45. $\dfrac{\dfrac{1}{x-5} - \dfrac{1}{x+4}}{\dfrac{2x^2-8}{x^2-x-20}}$

Use $(x-5)(x+4)$ as the LCD.

$$= \frac{(x-5)(x+4) \cdot \dfrac{1}{x-5} - \dfrac{1}{x+4} \cdot (x-5)(x+4)}{\dfrac{2x^2-8}{(x-5)(x+4)} \cdot (x-5)(x+4)}$$

Multiply each term by the LCD.

$$= \frac{x+4-x+5}{2x^2-8}$$

$$= \frac{9}{2x^2-8}$$

(Rational expressions)

46. $\quad -\sqrt{54x^3} = -\sqrt{9 \cdot 6x^3} = -3x\sqrt{6x}$

    (Radicals and rational exponents)

47. $\quad \sqrt[4]{32x^9y^4} = \sqrt[4]{2^5x^9y^4} = 2x^2y\,\sqrt[4]{2x}$

    (Radicals and rational exponents)

48. $\quad \sqrt[3]{\dfrac{12x^5y^4}{16xy^2}} = \sqrt[3]{\dfrac{6x^4y^2}{8}} = \dfrac{\sqrt[3]{6x^4y^2}}{\sqrt[3]{8}} = \dfrac{x\sqrt[3]{6xy^2}}{2}$

    (Radicals and rational exponents)

49. $\quad 32^{3/5} = \left(\sqrt[5]{32}\right)^3 = \left(\sqrt[5]{2^5}\right)^3 = 2^3 = 8$

    (Radicals and rational exponents)

50. $\quad 25^{-1/2} = \dfrac{1}{25^{1/2}}$ Use $x^{-n} = \dfrac{1}{x^n}$.

    $\qquad\quad\;\; = \dfrac{1}{\sqrt{25}}$ Use $x^{p/q} = \sqrt[q]{x^p}$.

    $\qquad\quad\;\; = \dfrac{1}{5}$

    (Radicals and rational exponents)

51. $\quad x^{\frac{1}{2}} \cdot x^{\frac{2}{3}} = x^{\frac{1}{2}+\frac{2}{3}}$ Use $x^m x^n = x^{m+n}$.

    $\qquad\quad\; = x^{\frac{3}{6}+\frac{4}{6}}$ To add the fractions, use the LCD 6.

    $\qquad\quad\; = x^{7/6}$

    (Radicals and rational exponents)

52. $\quad \dfrac{x^{\frac{2}{3}}y^{\frac{1}{2}}}{(xy)^{\frac{1}{3}}} = \dfrac{x^{\frac{2}{3}}y^{\frac{1}{2}}}{x^{\frac{1}{3}}y^{\frac{1}{3}}}$ Use $(xy)^n = x^n y^n$.

    $\qquad\quad = x^{\frac{2}{3}-\frac{1}{3}}y^{\frac{1}{2}-\frac{1}{3}}$ Use $\dfrac{x^m}{x^n} = x^{m-n}$.

    $\qquad\quad = x^{\frac{1}{3}}y^{\frac{3}{6}-\frac{2}{6}}$

$$= x^{1/3}y^{1/6}$$

(Radicals and rational exponents)

53. $\dfrac{3}{\sqrt{8}} = \dfrac{3}{2\sqrt{2}}$ Simplify $\sqrt{8} = 2\sqrt{2}$.

$$= \dfrac{3}{2\sqrt{2}} \cdot \dfrac{\sqrt{2}}{\sqrt{2}}$$ Multiply by $\dfrac{\sqrt{2}}{\sqrt{2}}$.

$$= \dfrac{3\sqrt{2}}{4}$$

(Radicals and rational exponents)

54. $\dfrac{\sqrt{5}-3}{4} \cdot \dfrac{\sqrt{5}+3}{\sqrt{5}+3}$ Multiply the numerator and denominator by the conjugate of the numerator.

$$= \dfrac{(\sqrt{5})^2 - (3)^2}{4(\sqrt{5}+3)}$$ Use $(x-y)(x+y) = x^2 - y^2$.

$$= \dfrac{5-9}{4(\sqrt{5}+3)}$$

$$= \dfrac{-4}{4(\sqrt{5}+3)}$$ Reduce by dividing out the common factor 4.

$$= \dfrac{-1}{\sqrt{5}+3}$$

(Radicals and rational exponents)

55. $2\sqrt{50} - 5\sqrt{8}$

$$= 2\sqrt{25 \cdot 2} - 5\sqrt{4 \cdot 2}$$ Simplify each radical.

$$= 2 \cdot 5\sqrt{2} - 5 \cdot 2\sqrt{2}$$

$$= 10\sqrt{2} - 10\sqrt{2}$$

$$= 0$$

(Radicals and rational exponents)

56. $4\sqrt[3]{16} + 2\sqrt[3]{54}$

$$= 4\sqrt[3]{8 \cdot 2} + 2\sqrt[3]{27 \cdot 2}$$ Simplify each radical.

$$= 4 \cdot 2\sqrt[3]{2} + 2 \cdot 3\sqrt[3]{2}$$

$$= 8\sqrt[3]{2} + 6\sqrt[3]{2}$$ Add like radicals.

$= 14 \sqrt[3]{2}$

(Radicals and rational exponents)

57. $3\sqrt{2}(5 - 2\sqrt{2})$

   $= 15\sqrt{2} - 6\sqrt{4}$ Use the distributive property.

   $= 15\sqrt{2} - 12$ $\sqrt{4} = 2$.

(Radicals and rational exponents)

58. $(\sqrt{5} + 2\sqrt{3})(3\sqrt{5} - 4\sqrt{3})$

   $= 3\sqrt{25} - 4\sqrt{15} + 6\sqrt{15} - 8\sqrt{9}$ Use the distributive property.

   $= 3(5) + 2\sqrt{15} - 8(3)$ Combine like radicals.

   $= 15 + 2\sqrt{15} - 24$ Combine like terms.

   $= -9 + 2\sqrt{15}$

(Radicals and rational exponents)

59. $(8 - 5i) - (3 - 4i)$

   $= 8 - 5i - 3 + 4i$ Distribute the subtraction sign.

   $= 5 - i$ Combine like terms.

(Complex numbers)

60. $(2 + 3i)(2 - 3i)$

   $= (2)^2 - (3i)^2$ Use $(x - y)(x + y) = x^2 - y^2$.

   $= 4 - 9i^2$

   $= 4 - 9(-1)$ Use $i^2 = -1$.

   $= 4 + 9 = 13$

(Complex numbers)

61. $\sqrt{-12} + 5\sqrt{-27}$

   $= \sqrt{-1 \cdot 4 \cdot 3} + 5\sqrt{-1 \cdot 9 \cdot 3}$ Factor each radicand.

   $= 2i\sqrt{3} + 5(3i)\sqrt{3}$ Use $\sqrt{-1} = i$.

   $= 2i\sqrt{3} + 15i\sqrt{3}$

$= 17i\sqrt{3}$ Add like terms.

(Complex numbers)

62. $\dfrac{5-8i}{2i} \cdot \dfrac{i}{i}$ Multiply by $\dfrac{i}{i}$.

$= \dfrac{5i - 8i^2}{2i^2}$

$= \dfrac{5i + 8}{-2}$ Use $i^2 = -1$.

$= -\dfrac{5i + 8}{2}$

(Complex numbers)

63. $\dfrac{6}{4-i} \cdot \dfrac{4+i}{4+i}$ Multiply the numerator and denominator by the conjugate of the denominator.

$= \dfrac{24 + 6i}{16 - i^2}$

$= \dfrac{24 + 6i}{16 + 1}$ Use $i^2 = -1$.

$= \dfrac{24 + 6i}{17}$

(Complex numbers)

Grade Yourself

Circle the question numbers that you had incorrect. Then indicate the number of questions you missed. If you answered more than three questions incorrectly, you will have to focus on that topic. If a topic has fewer than three questions and you had at least one wrong, we suggest you study that topic. Read your textbook or a review book or ask your teacher for help.

Subject: Basic Algebra Review

Topic	Question Numbers	Number Incorrect
The real numbers	1, 2, 3, 4, 5, 6, 7, 8, 9, 10, 11, 12	
Integer exponents and scientific notation	13, 14, 15, 16, 17	
Polynomials	18, 19, 20, 21, 22, 23, 24, 25, 26, 27, 28, 29	
Factoring	30, 31, 32, 33, 34, 35, 36, 37	
Rational expressions	38, 39, 40, 41, 42, 43, 44, 45	
Radicals and rational exponents	46, 47, 48, 49, 50, 51, 52, 53, 54, 55, 56, 57, 58	
Complex numbers	59, 60, 61, 62, 63	

Equations and Inequalities

2

Brief Yourself

This chapter contains questions about solving linear equations, linear inequalities, and quadratic equations and modeling with linear equations, solving other types of equations, and solving other types of inequalities. While many of these topics are covered in intermediate algebra, some of the problems are more involved than those at an intermediate level.

To solve a linear equation isolate the variable. Conditional equations are true for only some values of the variable, while identities are true for all defined values of the variable. If the process of solving an equation yields an equation with no variables and a true statement (such as $0 = 0$), the solution set is the set of all real numbers, written $\Re$ or $(-\infty, \infty)$. If the process of solving an equation yields an equation with no variables and a false statement (such as $0 = 1$), the solution set is empty, written $\varnothing$. When a solution to an equation produces 0 in the denominator of the original equation, that solution cannot be used in the solution set.

To solve word problems with modeling, read the problem carefully (several times, if necessary), and assign variables to quantities in a meaningful manner (d for distance, t for time, etc.). Memorize any required formulas or program them into your calculator as required by your instructor.

Linear inequalities are solved in the same manner as linear equations except when multiplying or dividing by a negative number (in which case the inequality symbol is reversed). Answers are often displayed on a number line graph and written in interval notation. In interval notation, a parenthesis indicates that the endpoint value *is not included* in the solution, while a bracket indicates that the end point *is included* in the solution.

To solve absolute value inequalities, use the following rules (must be memorized):

$|x| < a$ if and only if $-a < x < a$

$|x| > a$ if and only if $x < -a$ or $x > a$

Note that $<$ can be replaced by $\leq$ and $>$ can be replaced by $\geq$ and the rules are still valid.

Quadratic equations can be solved by factoring, using the square root property, completing the square, and/or the quadratic formula. To solve by factoring or the quadratic formula, the equation must be written in the form $ax^2 + bx + c = 0$ with zero on one side of the equation. The square root property works well with quadratic equations of the form $u^2 = c$ where the square root is taken of both sides to give $u = \pm\sqrt{c}$. To use completing the square, the coefficient of x^2 must equal 1, which may require dividing each term by the coefficient of x^2.

The quadratic formula must be memorized:

If $ax^2 + bx + c = 0$, then $x = \dfrac{-b \pm \sqrt{b^2 - 4ac}}{2a}$.

The discriminant, $b^2 - 4ac$, can be used to determine the nature of the solutions of a quadratic equation:

If $b^2 - 4ac < 0$, there are two complex solutions.

If $b^2 - 4ac = 0$, there is one real solution of multiplicity 2.

If $b^2 - 4ac > 0$, there are two distinct real solutions.

Other types of equations can be solved by techniques similar to those used for solving linear and quadratic equations. Some polynomial equations can be solved by factoring or by making an appropriate substitution. If an equation contains a variable raised to a rational exponent, isolate the variable and raise both sides of the equation to an appropriate power. An equation involving the absolute value of a variable can be solved using:

$|x| = a$ if and only if $x = a$ or $x = -a$.

Note that this will require solving *two* separate equations. Solutions should be checked in all these cases.

Quadratic and higher order polynomial inequalities and rational inequalities are solved using a number line divided into regions bounded by critical numbers. The critical numbers for polynomial and quadratic inequalities are found by factoring and setting each factor equal to 0. The critical numbers for rational inequalities are found by isolating 0 on one side of the inequality and then setting the numerator and denominator equal to 0.

Test Yourself

Solve each equation.

1. $6x - 4 = 2x + 6$

2. $3(x + 5) - 5 = 2(3x - 4)$

3. $-4(2x + 5) + 10 = 2(10 - 4x) - 50$

4. $\dfrac{60 - 4k}{3} = \dfrac{5k + 6}{4} + 2$

5. $\dfrac{1}{2}(4x + 2) = 4x - 2(x - 3) - 5$

6. $8 - \dfrac{13}{x} = 2 + \dfrac{5}{x}$

7. $\dfrac{1}{x - 2} + \dfrac{3}{x + 3} = \dfrac{4}{x^2 + x - 6}$

8. $\dfrac{1}{(x - 3)(x - 2)} = \dfrac{1}{x - 3} + \dfrac{2}{x - 2}$

9. $\dfrac{1}{y} + \dfrac{3}{y - 2} = 0$

10. Solve for x: $2x - 3y = 6$

11. Solve for R_1: $\dfrac{1}{R} = \dfrac{1}{R_1} - \dfrac{1}{R_2}$

12. Solve for h: $A = 2\pi rh + \pi r^2$

Write an algebraic expression for the verbal description below. Indicate what each variable represents.

13. Your maximum target heart rate is 85 percent of the difference between 220 and your age.

14. Find the maximum target heart rate for a 20-year-old.

15. The price of a luxury automobile has been discounted 22.5 percent for a year-end clearance sale. The sale price is $34,875. Find the original price of the car.

16. Suppose you are taking a course with four tests that are 100 points each and a final test that is 200 points. To get an A in the course you must have an average of at least 90 percent on the five tests. Your scores on the first four tests were 87, 90, 92, and 84. What must you score on the fifth test to get an A for the course?

17. The chemistry professor has 80 gallons of a mixture with a concentration of 40 percent. How much of the mixture should be withdrawn and replaced by 100 percent concentrate to bring the mixture up to 75 percent concentration?

18. You plan to invest $8,000 in two funds paying 6 percent and 8 1/2 percent simple interest. (There is more risk in the 8 1/2 percent fund.) Your goal is to obtain a total annual interest income of $542.50 from the investments. What is the smallest amount you can invest in the 8 1/2 percent fund in order to meet your objective?

19. A small company has fixed costs of $1,200 per month and variable costs of $6.50 per unit produced. If each unit sells for $12.50, how many units must be sold to break even? How many units must be sold to make a profit of $2,000?

20. On the first part of a 774-mile trip, Pat averaged 68 miles per hour. She averaged only 64 miles per hour on the last part of the trip. Find the amount of time at each of the speeds if the total time was 11 hours and 45 minutes.

Solve each inequality, write solution sets using interval notation and graph each solution set on a number line.

21. $4x - 8 < 4$

22. $4 - 2y \leq 3$

23. $-6 < 4 - 2x < 6$

24. $\frac{1}{2}x + 3 \leq \frac{1}{3}x + \frac{5}{2}$ or $3(x + 1) > 4$

25. $0.8(x + 2) - 0.3(2x + 1) > 0.2(x + 5)$

26. $|2x - 1| < 7$

27. $\left|x - \frac{8}{3}\right| + \frac{1}{3} \geq 1$

28. $|4(m + 2)| < -3$

29. $|2p + 6| > -1$

30. $|6 - 2x| - 4 \leq -1$

31. Determine whether the values of x are solutions of the inequality $-2 < \frac{4 - x}{3} \leq 2$.

    (a) $x = 0$

    (b) $x = -2$

    (c) $x = 10$

    (d) $x = \sqrt{10}$

32. The revenue for selling x units of a product is $R = 118.50x$. The cost of producing x units is $C = 85x + 675$. To obtain a profit, the revenue must be greater than the cost. For what values of x will this product return a profit?

Solve each equation.

33. $3x^2 = 24$

34. $3x^2 = 24x$

35. $2x^2 - 7x - 15 = 0$

36. $(x + 2)(2x + 3) = 10$

37. $\frac{1}{3}y^2 + \frac{1}{2}y = \frac{2}{3}$

38. $\left(x - \frac{1}{3}\right)^2 = -\frac{2}{3}$

39. $\frac{1}{2}p^2 + p - \frac{1}{5} = 0$

40. $\frac{1}{m^2} - \frac{4}{m} + 1 = 0$

41. Use the discriminant to determine the number of solutions to $x^2 + 7x + 10 = 0$.

42. Use the discriminant to determine the number of solutions to $y^2 - 3y + 9 = 0$.

43. Solve $2x^2 + 16x + 28 = 0$ by completing the square.

Find all solutions.

44. $x^4 - 16 = 0$

45. $x^3 - 3x^2 + 6x - 18 = 0$

46. $x^4 - 3x^2 - 4 = 0$

47. $8 + \dfrac{10}{p} - \dfrac{3}{p^2} = 0$

48. $5x + 2\sqrt{x} = 1$

49. $x^{2/3} - x^{1/3} - 6 = 0$

50. $\sqrt{x+3} - x = 1$

51. $\sqrt{x} - \sqrt{x-16} = 1$

52. $(x-4)^{2/3} = 16$

53. $\dfrac{1}{x} - \dfrac{1}{x+1} = 4$

54. $\dfrac{-2}{y+2} + \dfrac{y}{y+4} = \dfrac{2y}{y^2 + 6y + 8}$

55. $|3x + 1| = 5$

56. $|2x - 1| = \dfrac{1}{3}|2x + 1|$

Solve each inequality.

57. $x^2 - x - 6 \le 0$

58. $2x^2 > 5x + 3$

59. $(x+1)(x-2)(x+4) \ge 0$

60. $x(x-2)^2(x+1) < 0$

61. $\dfrac{4}{x-1} \le 2$

62. $\dfrac{2}{x+5} > \dfrac{1}{2x+3}$

Check Yourself

1.
$$6x - 4 = 2x + 6$$

$$6x - 4 + (-2x + 4) = 2x + 6 + (-2x + 4)$$ Isolate x.

$$4x = 10$$

$$\frac{4x}{4} = \frac{10}{4}$$ Divide each side by 4.

$$x = \frac{5}{2}$$ Reduce.

(Linear equations)

2.
$$3(x + 5) - 5 = 2(3x - 4)$$

$$3x + 15 - 5 = 6x - 8$$ Use the distributive property.

$$3x + 10 = 6x - 8$$ Simplify each side.

$$3x + 10 + (-6x - 10) = 6x - 8 + (-6x - 10)$$ Isolate x.

$$-3x = -18$$

$$x = 6$$

(Linear equations)

3. $$-4(2x+5) + 10 = 2(10-4x) - 50$$

$$-8x - 20 + 10 = 20 - 8x - 50 \qquad \text{Use the distributive property.}$$

$$-8x - 10 = -8x - 30 \qquad \text{Simplify each side.}$$

$$-8x - 10 + (8x + 10) = -8x - 30 + (8x + 10)$$

$$0 = -20$$

There are no remaining variables and $0 \neq -20$, therefore the solution set is empty, $\varnothing$.

(Linear equations)

4. $$\frac{60-4k}{3} = \frac{5k+6}{4} + 2$$

$$12 \cdot \frac{60-4k}{3} = 12 \cdot \frac{5k+6}{4} + 12 \cdot 2 \qquad \text{Multiply each term by 12, the LCD.}$$

$$4(60-4k) = 3(5k+6) + 12(2) \qquad \text{Cancel.}$$

$$240 - 16k = 15k + 18 + 24 \qquad \text{Use the distributive property.}$$

$$240 - 16k = 15k + 42 \qquad \text{Simplify each side.}$$

$$240 - 16k + (-240 - 15k) = 15k + 42 + (-240 - 15k) \qquad \text{Isolate } k.$$

$$-31k = -198$$

$$k = \frac{-198}{-31} = \frac{198}{31}$$

(Linear equations)

5. $$\frac{1}{2}(4x+2) = 4x - 2(x-3) - 5$$

$$2x + 1 = 4x - 2x + 6 - 5 \qquad \text{Multiply.}$$

$$2x + 1 = 2x + 1 \qquad \text{Simplify each side.}$$

$$0 = 0$$

Since there are no remaining variables and $0 = 0$ is a true statement, the solution set is all real numbers, $\Re$ or $(-\infty, \infty)$.

(Linear equations)

6.
$$8 - \frac{13}{x} = 2 + \frac{5}{x}$$

$$x \cdot 8 - x \cdot \frac{13}{x} = 2x + \frac{5}{x} \cdot x$$ Multiply each term by x, the LCD, $(x-2)(x+3)$.

$$8x - 13 = 2x + 5$$ Cancel.

$$6x = 18$$

$$x = 3$$

(Linear equations)

7.
$$\frac{1}{x-2} + \frac{3}{x+3} = \frac{4}{x^2 + x - 6}$$

$$(x-2)(x+3) \cdot \frac{1}{x-2} + (x-2)(x+3) \cdot \frac{3}{x+3} = \frac{4}{(x-2)(x+3)} \cdot (x-2)(x+3)$$

     Multiply each term by the LCD.

$$x + 3 + 3(x-2) = 4$$ Cancel.

$$x + 3 + 3x - 6 = 4$$ Multiply.

$$4x - 3 = 4$$ Simplify the left side.

$$4x = 7$$

$$x = \frac{7}{4}$$

(Linear equations)

8.
$$\frac{1}{(x-3)(x-2)} = \frac{1}{x-3} + \frac{2}{x-2}$$

$$(x-3)(x-2) \cdot \frac{1}{(x-3)(x-2)} = (x-3)(x-2) \cdot \frac{1}{x-3} + (x-3)(x-2) \cdot \frac{2}{x-2}$$

$$1 = x - 2 + 2x - 6$$

$$1 = 3x - 8$$

$$9 = 3x$$

$$3 = x$$

But, if $x = 3$, the denominator in the original equation equals 0. Therefore, the solution set is empty, $\varnothing$.

(Linear equations)

9.
$$\frac{1}{y} + \frac{3}{y-2} = 0$$

$$y(y-2)\cdot\frac{1}{y} + y(y-2)\cdot\frac{3}{y-2} = y(y-2)(0)$$ Multiply each term by the LCD, $y(y-2)$.

$$y - 2 + 3y = 0$$ Cancel.

$$4y = 2$$ Solve for y.

$$y = \frac{1}{2}$$

(Linear equations)

10. $2x - 3y = 6$

$$2x = 3y + 6$$ Add $3y$ to each side.

$$\frac{2x}{2} = \frac{3y+6}{2}$$ Divide each side by 2.

$$x = \frac{3}{2}y + 3$$

(Linear equations)

11.
$$\frac{1}{R} = \frac{1}{R_1} - \frac{1}{R_2}$$

$$RR_1R_2\cdot\frac{1}{R} = RR_1R_2\cdot\frac{1}{R_1} - RR_1R_2\cdot\frac{1}{R_2}$$ Multiply each term by the LCD.

$$R_1R_2 = RR_2 - RR_1$$ Cancel.

$$R_1R_2 + RR_1 = RR_2$$ Add RR_1 to both sides.

$$R_1(R_2 + R) = RR_2$$ Common factor R_1.

$$R_1 = \frac{RR_2}{R_2 + R}$$ Divide each side by $R_2 - R$.

(Linear equations)

12. $A = 2\pi rh + \pi r^2$

$$A - \pi r^2 = 2\pi rh$$ Isolate h.

$$\frac{A - \pi r^2}{2\pi r} = h$$ Divide each side by $2\pi r$, the coefficient of h.

(Linear equations)

13. $M = 0.85(220 - A)$

where M = maximum target heart rate and A = age

(Modeling with linear equations)

14. $M = 0.85(220 - 20)$ Substitute $A = 20$.

$M = 170$

(Modeling with linear equations)

15.   Let p = the original price of the car. Write an equation to represent the 22.5 percent discount:

$100\% \, p - 22.5\% \, p = 34,875$

$77.5\% \, p = 34,875$

$0.775 \, p = 34875$ Convert the percent to a decimal.

$p = \dfrac{34875}{0.775}$

$p = 45,000$

Thus, the original price of the car was $45,000.

(Modeling with linear equations)

16.   Let s = score on the fifth test. There is a total of 600 points available $(4 \times 100 + 1 \times 200)$. To average 90 percent, you need 90 percent of 600 points or 540 points.

$87 + 90 + 92 + 84 + s = 540$

$353 + s = 540$

$s = 187$

Thus, you must score at least 187 points on the fifth test.

(Modeling with linear equations)

17.   Let x = number of gallons withdrawn, which is also the amount to be added to keep a total of 80 gallons. Then,

$40\%(80 - x) + 100\%x = 75\%(80)$

$.4(80 - x) + x = .75(80)$ Convert the percents to decimals.

$40(80 - x) + 100x = 75(80)$ Multiply each term by 100 to clear the decimals.

$3200 - 40x + 100x = 6000$

$60x = 2800$

$x = 46\dfrac{2}{3}$

Therefore, $46\frac{2}{3}$ gallons should be withdrawn.

(Modeling with linear equations)

18. Let x = amount invested at 6 percent.

$800 - x$ = amount invested at $8\frac{1}{2}$ percent.

Simple interest is found by multiplying the amount invested times rate times the time (1 year).

$6\%x + 8\frac{1}{2}\%(8000 - x) = \$542.50$

$0.06x + 0.085(800 - x) = 542.50$ Convert the percents to decimals.

$60x + 85(8000 - x) = 542500$ Multiply by 1,000 to clear decimals.

$60x + 680000 - 85x = 542500$

$-25x = -137500$

$x = 5500$

Therefore, the smallest amount that can be invested at $8\frac{1}{2}$ percent is $8,000 - 5,500 = \$2,500$.

(Modeling with linear equations)

19. The break-even point occurs when revenue equals cost. Costs consist of fixed costs plus variable costs. Let x = number of units produced.

$12.50x = 1200 + 6.50x$

$6x = 1200$

$x = 200$

Thus, 200 units must be sold to break even.

Since profit = revenue – cost,

$2000 = 12.50x - (1200 + 6.50x)$

$2000 = 6x - 1200$

$3200 = 6x$

$533.33 = x$

Assuming x must be an integer, this means 534 units must be sold to make a profit of \$2,000.

(Modeling with linear equations)

20. Let t = time in hours for the first part of the trip. Then $11.75 - t$ = time in hours for the last part of the trip.

Use the formula $d = rt$ to write:

$$774 = 68t + (11.75 - t)64$$

$$774 = 68t + 752 - 64t$$

$$774 = 4t + 752$$

$$22 = 4t$$

$$5.5 = t$$

Therefore, the first part of the trip took 5.5 hours or 5 hours 30 minutes, and the second part took $11.75 - 5.5 = 6.25$ hours or 6 hours 15 minutes.

(Modeling with linear equations)

21. $4x - 8 < 4$

        $4x < 12$ Add 8 to each side.

         $x < 3$ Divide each side by 4.

interval notation: $(-\infty, 3)$

graph:

(Linear inequalities)

22. $4 - 2y \le 3$

       $-2y \le -1$ Subtract 4 from each side.

         $y \ge \dfrac{1}{2}$ Divide each side by –2. Reverse the inequality symbol.

interval notation: $\left[\dfrac{1}{2}, \infty \right)$

graph:

(Linear inequalities)

23. $-6 < 4 - 2x < 6$

      $-10 < -2x < 2$ Subtract 4 from each part.

      $\dfrac{-10}{-2} > \dfrac{-2x}{-2} > \dfrac{2}{-2}$ Divide each part by – 2. Reverse the inequality symbols.

      $5 > x > -1$

interval notation: $(-1, 5)$

graph:

$$-1 \qquad 5$$

(Linear inequalities)

24.   Solve each inequality:

$$\frac{1}{2}x + 3 \le \frac{1}{3}x + \frac{5}{2}$$

$$6 \cdot \frac{1}{2}x + 6 \cdot 3 \le 6 \cdot \frac{1}{3}x + 6 \cdot \frac{5}{2} \qquad\qquad \text{Multiply each term by the LCD, 6.}$$

$$3x + 18 \le 2x + 15$$

$$x \le -3$$

OR:

$$3(x + 1) > 4$$

$$3x + 3 > 4$$

$$3x > 1$$

$$x > \frac{1}{3}$$

Thus, $x \le -3$ or $x > \frac{1}{3}$.

interval notation: $(-\infty, -3] \cup \left(\frac{1}{3}, \infty\right)$

graph:

$$-3 \qquad 1/3$$

(Linear inequalities)

25. $0.8(x + 2) - 0.3(2x + 1) > 0.2(x + 5)$

$$8(x + 2) - 3(2x + 1) > 2(x + 5) \qquad\qquad \text{Multiply each term by 10.}$$

$$8x + 16 - 6x - 3 > 2x + 10$$

$$2x + 13 > 2x + 10$$

$$0 > -3$$

There are no variables and $0 > -3$ is a true statement. The solution set is all real numbers.

interval notation: $(-\infty, \infty)$

graph:

(Linear inequalities)

26. $|2x - 1| < 7$

     $-7 < 2x - 1 < 7$ Use: $|x| < a \Leftrightarrow -a < x < a$.

     $-6 < 2x < 8$ Add 1 to each part.

     $-3 < x < 4$ Divide each part by 2.

     interval notation: $(-3, 4)$

     graph:

(Linear inequalities)

27. $\left| x - \dfrac{8}{3} \right| + \dfrac{1}{3} \geq 1$

     $\left| x - \dfrac{8}{3} \right| \geq \dfrac{2}{3}$ Isolate the absolute value by subtracting $\dfrac{1}{3}$ from each side.

     $x - \dfrac{8}{3} \geq \dfrac{2}{3}$ or $x - \dfrac{8}{3} \leq -\dfrac{2}{3}$ Use: $|x| \geq a \Leftrightarrow x \geq a$ or $x \leq -a$.

     $x \geq \dfrac{10}{3}$ or $x \leq 2$ Solve each inequality.

     interval notation: $\left(-\infty, 2 \right] \cup \left[\dfrac{10}{3}, \infty \right)$

     graph:

(Linear inequalities)

28. $|4(m + 2)| < -3$

Although $|x| < a \Leftrightarrow -a < x < a$, it is quicker to realize that no value of m will make the absolute value less than any negative number. Thus, the solution set is empty, $\varnothing$.

(Linear inequalities)

29. $|2p + 6| > -1$

Although $|x| > a \Leftrightarrow x > a$ or $x < -a$, it is quicker to realize that any value of p will make the absolute value greater than any negative number. Thus, the solution set is $(-\infty, \infty)$.

graph:

0

(Linear inequalities)

30. $|6 - 2x| - 4 \leq -1$

Do not use the shortcut from problem 28 because the absolute value has not been isolated.

$|6 - 2x| \leq 3$

$-3 \leq 6 - 2x \leq 3$	Use $\lvert x \rvert \leq a \Leftrightarrow -a \leq x \leq a$.
$-9 \leq -2x \leq -3$	Add –6 to each part.
$\dfrac{9}{2} \geq x \geq \dfrac{3}{2}$	Divide each part by –2 and reverse the inequality symbols.

interval notation: $\left[\dfrac{3}{2}, \dfrac{9}{2}\right]$

graph:

3/2   9/2

(Linear inequalities)

31. Solve the given inequality:

$$-2 < \frac{4-x}{3} \leq 2$$

$3(-2) < 3 \cdot \dfrac{4-x}{3} \leq 3(2)$	Multiply each part by 3.
$-6 < 4 - x \leq 6$	
$-10 < -x \leq 2$	Add –4 to each part.
$10 > x \geq -2$	Divide each part by –1 and reverse the inequality symbols.

Now, since $-2 \leq x < 10$, values 0 and –2 *are* solutions of the original inequality. 10 is *not* a solution since x is less than 10. Since $\sqrt{10}$ is between $3 = \sqrt{9}$ and $4 = \sqrt{16}$, $\sqrt{10}$ *is* a solution.

(Linear inequalities)

32.

$R > C$	Set up an inequality.
$118.50x > 85x + 675$	Use the given revenue and cost equations.
$33.5x > 675$	
$x > 20.15$	

Assume units must be whole numbers, so $x \geq 21$ to return a profit.

(Linear inequalities)

33. $3x^2 = 24$

      $x^2 = 8$ Divide each side by 8.

      $x = \pm\sqrt{8}$ Take the square root of each side, using $\pm$ on the right side.

      $x = \pm 2\sqrt{2}$ Simplify the radical.

(Quadratic equations)

34. $3x^2 = 24x$

$3x^2 - 24x = 0$

$3x(x-8) = 0$ Factor.

$3x = 0$ or $x - 8 = 0$ Set each factor equal to 0.

  $x = 0$ or $x = 8$ Solve.

Note that if you had divided each side by $3x$ to give $x = 8$, the solution $x = 0$ would have been lost.

(Quadratic equations)

35. $2x^2 - 7x - 15 = 0$

  $(2x + 3)(x - 5) = 0$ Factor.

  $2x + 3 = 0$ or $x - 5 = 0$ Set each factor equal to 0.

      $x = -\dfrac{3}{2}$ or $x = 5$ Solve.

(Quadratic equations)

36. $(x + 2)(2x + 3) = 10$

    $2x^2 + 7x + 6 = 10$

    $2x^2 + 7x - 4 = 0$ Subtract 10 from each side.

  $(2x - 1)(x + 4) = 0$ Factor.

  $2x - 1 = 0$ or $x + 4 = 0$ Set each factor equal to 0.

      $x = \dfrac{1}{2}$ or $x = -4$

Note that to set each factor equal to 0, you must first have 0 isolated on one side of the equation.

(Quadratic equations)

37. $\dfrac{1}{3}y^2 + \dfrac{1}{2}y = \dfrac{2}{3}$

     $6 \cdot \dfrac{1}{3}y^2 + 6 \cdot \dfrac{1}{2}y = 6 \cdot \dfrac{2}{3}$ Multiply each term by the LCD 6.

     $2y^2 + 3y = 4$

     $2y^2 + 3y - 4 = 0$ Subtract 4 from each side.

$$y = \dfrac{-3 \pm \sqrt{(3)^2 - 4(2)(-4)}}{2(2)}$$ Use the quadratic formula.

$$y = \dfrac{-3 \pm \sqrt{41}}{4}$$ Simplify the radicand.

(Quadratic equations)

38. $\left(x - \dfrac{1}{3}\right)^2 = -\dfrac{2}{3}$

     $\sqrt{\left(x - \dfrac{1}{3}\right)^2} = \pm\sqrt{-\dfrac{2}{3}}$ Take the square root of each side.

     $x - \dfrac{1}{3} = \pm\dfrac{i\sqrt{2}}{\sqrt{3}}$ Use $\sqrt{-\dfrac{2}{3}} = \sqrt{-1} \cdot \dfrac{\sqrt{2}}{\sqrt{3}} = \dfrac{i\sqrt{2}}{\sqrt{3}}$.

     $x = \dfrac{1}{3} \pm \dfrac{i\sqrt{2}}{\sqrt{3}}$ Add $\dfrac{1}{3}$ to each side.

     $x = \dfrac{1}{3} \pm \dfrac{i\sqrt{2}}{\sqrt{3}} \cdot \dfrac{\sqrt{3}}{\sqrt{3}}$ Rationalize the denominator.

     $x = \dfrac{1 \pm i\sqrt{6}}{3}$

(Quadratic equations)

39. $\dfrac{1}{2}p^2 + p - \dfrac{1}{5} = 0$

     $10 \cdot \dfrac{1}{2}p^2 + 10p - 10 \cdot \dfrac{1}{5} = 10 \cdot 0$ Multiply each term by 10.

     $5p^2 + 10p - 2 = 0$

$$p = \frac{-10 \pm \sqrt{(10)^2 - 4(5)(-2)}}{2(5)}$$

Use the quadratic formula.

$$p = \frac{-10 \pm \sqrt{140}}{10}$$

$$p = \frac{-10 \pm 2\sqrt{35}}{10}$$

$\sqrt{140} = \sqrt{4 \cdot 35} = 2\sqrt{35}$.

$$p = \frac{2(-5 \pm \sqrt{35})}{10}$$

Common factor 2.

$$p = \frac{-5 \pm \sqrt{35}}{5}$$

Reduce $\frac{2}{10} = \frac{1}{5}$.

(Quadratic equations)

40. $$\frac{1}{m^2} - \frac{4}{m} + 1 = 0$$

$$m^2 \cdot \frac{1}{m^2} - m^2 \cdot \frac{4}{m} + m^2 \cdot 1 = m^2 \cdot 0$$

Multiply each term by the LCD m^2.

$$1 - 4m + m^2 = 0 \text{ or } m^2 - 4m + 1 = 0$$

Write the equation in $ax^2 + bx + c = 0$ form.

$$m = \frac{-(-4) \pm \sqrt{(-4)^2 - 4(1)(1)}}{2(1)}$$

Use the quadratic formula.

$$m = \frac{4 \pm \sqrt{12}}{2}$$

Simplify.

$$m = \frac{4 \pm 2\sqrt{3}}{2}$$

$\sqrt{12} = \sqrt{4}\sqrt{3} = 2\sqrt{3}$.

$$m = \frac{2(2 \pm \sqrt{3})}{2}$$

Factor the numerator.

$$m = 2 \pm \sqrt{3}$$

Reduce.

(Quadratic equations)

41. In $x^2 + 7x + 10 = 0$, $a = 1, b = 7, c = 10$ so the discriminant $b^2 - 4ac = 7^2 - 4(1)(10) = 49 - 40 = 9$. Since the discriminant is greater than 0, there are two distinct real solutions.

(Quadratic equations)

42. In $y^2 - 3y + 9 = 0$, $a = 1, b = -3, c = 9$. $b^2 - 4ac = (-3)^2 - 4(1)(9) = 9 - 36 = -27$. Since the discriminant is less than 0, there are two complex solutions.

(Quadratic equations)

43. $2x^2 + 16x + 28 = 0$

$x^2 + 8x + 14 = 0$ Divide each term by the coefficient of x^2, 2.

$x^2 + 8x = -14$ Subtract 14 from each side.

$x^2 + 8x + 16 = -14 + 16$ Add $\left(\frac{1}{2} \cdot 8\right)^2 = 4^2 = 16$ to each side.

$(x+4)^2 = 2$ Factor the left side; simplify the right side.

$x + 4 = \pm\sqrt{2}$ Use $x^2 = a \Leftrightarrow x = \pm\sqrt{a}$.

$x = -4 \pm \sqrt{2}$ Isolate x.

(Quadratic equations)

44. $x^4 - 16 = 0$

$(x^2 - 4)(x^2 + 4) = 0$ Factor using $x^2 - y^2 = (x-y)(x+y)$.

$(x - 2)(x + 2)(x^2 + 4) = 0$ Factor.

$x - 2 = 0$ or $x + 2 = 0$ or $x^2 = -4$

$x = 2$ or $x = -2$ or $x = \pm\sqrt{-4}$ Solve each equation.

$x = \pm 2i$

The solution set is $\{-2, 2, -2i, 2i\}$.

(Other types of equations)

45. $x^3 - 3x^2 + 6x - 18 = 0$

$x^2(x - 3) + 6(x - 3) = 0$ Factor by grouping.

$(x - 3)(x^2 + 6) = 0$

$x - 3 = 0$ $x^2 + 6 = 0$ Set each factor equal to 0.

$x = 3$ $x = \pm i\sqrt{6}$ Solve.

The solution set is $\{3, \pm i\sqrt{6}\}$.

(Other types of equations)

46. $x^4 - 3x^2 - 4 = 0$

$(x^2 - 4)(x^2 + 1) = 0$ Factor by trial and error.

$$(x-2)(x+2)(x^2+1) = 0 \qquad \text{Factor using } x^2 - y^2 = (x-y)(x+y).$$

$$x-2 = 0 \quad x+2 = 0 \quad x^2+1 = 0 \qquad \text{Set each factor equal to 0.}$$

$$x = 2 \qquad x = -2 \qquad x = \pm i \qquad \text{Solve each equation.}$$

The solution set $\{\pm 2, \pm i\}$.

(Other types of equations)

47.
$$8 + \frac{10}{p} - \frac{3}{p^2} = 0$$

$$p^2 \cdot 8 + p^2 \cdot \frac{10}{p} - p^2 \cdot \frac{3}{p^2} = p^2 \cdot 0 \qquad \text{Multiply each term by the LCD, } p^2.$$

$$8p^2 + 10p - 3 = 0 \qquad \text{Cancel.}$$

$$(4p-1)(2p+3) = 0 \qquad \text{Factor by trial and error.}$$

$$4p-1 = 0 \qquad\qquad 2p+3 = 0 \qquad \text{Set each factor equal to 0.}$$

$$p = \frac{1}{4} \qquad\qquad p = -\frac{3}{2} \qquad \text{Solve each equation.}$$

The solution set is $\left\{-\dfrac{3}{2}, \dfrac{1}{4}\right\}$.

(Other types of equations)

48. $5x + 2\sqrt{x} = 1$

This equation is quadratic in form. Make the substitution $u = \sqrt{x}$, so $u^2 = x$:

$$5u^2 + 2u = 1$$

$$5u^2 + 2u - 1 = 0 \qquad \text{Write the equation in standard form.}$$

$$u = \frac{-2 \pm \sqrt{(2)^2 - 4(5)(-1)}}{2(5)} \qquad \text{Use the quadratic formula.}$$

$$u = \frac{-2 \pm \sqrt{24}}{10} \qquad \text{Simplify the radicand.}$$

$$u = \frac{-2 \pm 2\sqrt{6}}{10} \qquad\qquad \sqrt{24} = \sqrt{4}\sqrt{6} = 2\sqrt{6}.$$

$$u = \frac{2(-1 \pm \sqrt{6})}{10} \qquad \text{Factor the numerator.}$$

$$u = \frac{-1 \pm \sqrt{6}}{5} \qquad\qquad \text{Reduce.}$$

Since $u = \sqrt{x}$, complete the problem by returning to x values:

$$\sqrt{x} = \frac{-1 + \sqrt{6}}{5} \qquad \text{or} \qquad \sqrt{x} = \frac{-1 - \sqrt{6}}{5}$$

$$x = \left(\frac{-1 + \sqrt{6}}{5}\right)^2 \qquad \text{or} \qquad x = \left(\frac{-1 - \sqrt{6}}{5}\right)^2 \qquad \text{Square both sides.}$$

$$x = \frac{1 - 2\sqrt{6} + 6}{25} \qquad\qquad x = \frac{1 + 2\sqrt{6} + 6}{25}$$

$$x = \frac{7 - 2\sqrt{6}}{25} \qquad\qquad x = \frac{7 + 2\sqrt{6}}{25}$$

(Other types of equations)

49. $\quad x^{2/3} - x^{1/3} - 6 = 0$

This equation is quadratic in form. Let $u = x^{1/3}$ so $u^2 = x^{2/3}$.

$$u^2 - u - 6 = 0$$

$$(u - 3)(u + 2) = 0 \qquad\qquad \text{Factor.}$$

$$u - 3 = 0 \text{ or } u + 2 = 0 \qquad\qquad \text{Set each factor equal to 0.}$$

$$u = 3 \text{ or } \quad u = -2 \qquad\qquad \text{Solve for } u.$$

Now replace u with $x^{1/3}$.

$$x^{1/3} = 3 \text{ or } x^{1/3} = -2$$

$$\left(x^{1/3}\right)^3 = (3)^3 \qquad\qquad \left(x^{1/3}\right)^3 = (-2)^3 \qquad \text{Cube each side.}$$

$$x = 27 \qquad\qquad\qquad x = -8$$

Both solutions check, so the solution set is $\{-8, 27\}$.

(Other types of equations)

50. $\quad \sqrt{x + 3} - x = 1$

$$\sqrt{x + 3} = 1 + x \qquad\qquad \text{Isolate the radical.}$$

$$\left(\sqrt{x + 3}\right)^2 = (1 + x)^2 \qquad\qquad \text{Square both sides.}$$

$$x + 3 = 1 + 2x + x^2$$

$$0 = -2 + x + x^2$$

$$x^2 + x - 2 = 0$$

$$(x + 2)(x - 1) = 0 \qquad \text{Factor.}$$

$$x + 2 = 0 \text{ or } x - 1 = 0$$

$$x = -2 \text{ or } x = 1$$

Check each solution:

$$\sqrt{-2 + 3} - (-2) = 1 \qquad \text{Check } x = -2.$$

$$\sqrt{1} + 2 \neq 1$$

$$\sqrt{1 + 3} - (1) = 1 \qquad \text{Check } x = 1.$$

$$\sqrt{4} - 1 = 1$$

$$2 - 1 = 1$$

Only $x = 1$ checks so the solution set is $\{1\}$.

(Other types of equations)

51. $\quad \sqrt{x} - \sqrt{x - 16} = 1$

$$\sqrt{x} = 1 + \sqrt{x - 16} \qquad \text{Isolate one of the radicals.}$$

$$\left(\sqrt{x}\right)^2 = \left(1 + \sqrt{x - 16}\right)^2 \qquad \text{Square both sides.}$$

$$x = 1 + 2\sqrt{x - 16} + x - 16$$

$$x = -15 + 2\sqrt{x - 16} + x \qquad \text{Combine similar terms.}$$

$$15 = 2\sqrt{x - 16} \qquad \text{Isolate the radical.}$$

$$\frac{15}{2} = \sqrt{x - 16}$$

$$\left(\frac{15}{2}\right)^2 = \left(\sqrt{x - 16}\right)^2 \qquad \text{Square both sides.}$$

$$\frac{225}{4} = x - 16$$

$$\frac{289}{4} = x \qquad\qquad \frac{225}{4} + \frac{16}{1} = \frac{225}{4} + \frac{64}{4} = \frac{289}{4}.$$

Check the solution:

$$\sqrt{\frac{289}{4}} - \sqrt{\frac{289}{4} - 16} = 1$$

$$\sqrt{\frac{289}{4}} - \sqrt{\frac{289}{4} - \frac{64}{4}} = 1$$

$$\frac{17}{2} - \sqrt{\frac{225}{4}} = 1$$

$$\frac{17}{2} - \frac{15}{2} = 1$$

$$\frac{2}{2} = 1$$

The solution checks, therefore, the solution set is $\left\{\frac{289}{4}\right\}$.

(Other types of equations)

52. $(x-4)^{2/3} = 16$

$\left[(x-4)^{2/3}\right]^{3/2} = (16)^{3/2}$ Raise each side to the 3/2 power.

$x - 4 = 64$ $16^{3/2} = (16^{1/2})^3 = (4)^3 = 64$.

$x = 68$

The solution checks, so the solution set is {68}.

(Other types of equations)

53. $$\frac{1}{x} - \frac{1}{x+1} = 4$$

$x(x+1) \cdot \frac{1}{x} - x(x+1) \cdot \frac{1}{x+1} = x(x+1)4$ Multiply each term by the LCD, $x(x+1)$.

$$x + 1 - x = 4x(x+1)$$

$$1 = 4x^2 + 4x$$

$4x^2 + 4x - 1 = 0$ Write the equation in standard form.

$x = \dfrac{-4 \pm \sqrt{(4)^2 - 4(4)(-1)}}{2(4)}$ Use the quadratic formula.

$x = \dfrac{-4 \pm \sqrt{32}}{8}$

$$x = \frac{-4 \pm 4\sqrt{2}}{8}$$

$$\sqrt{32} = \sqrt{16}\sqrt{2} = 4\sqrt{2}.$$

$$x = \frac{-1 \pm \sqrt{2}}{2}$$

Factor out 4 from the numerator and reduce.

The solution set is $\left\{ \dfrac{-1 \pm \sqrt{2}}{2} \right\}$.

(Other types of equations)

54. $\dfrac{-2}{y+2} + \dfrac{y}{y+4} = \dfrac{2y}{y^2 + 6y + 8}$

$$(y+2)(y+4) \cdot \frac{-2}{y+2} + (y+2)(y+4) \cdot \frac{y}{y+4} = (y+2)(y+4) \cdot \frac{2y}{(y+2)(y+4)}$$

$-2(y+4) + y(y+2) = 2y$ Multiply.

$-2y - 8 + y^2 + 2y = 2y$ Combine similar terms.

$y^2 - 2y - 8 = 0$

$(y-4)(y+2) = 0$ Set each factor equal to 0 and solve.

$y = 4$ or $y = -2$

Note that if $y = -2$, the denominator $y + 2 = 0$, so -2 cannot be a solution. The solution 4 checks, so the solution set is {4}.

(Other types of equations)

55. $|3x + 1| = 5$

$3x + 1 = 5$ or $3x + 1 = -5$ Use $|x| = a \Leftrightarrow x = a$ or $x = -a$.

$x = \dfrac{4}{3}$ or $x = -2$ Solve each equation.

The solution set is $\{-2, \dfrac{4}{3}\}$.

(Other types of equations)

56. $|2x - 1| = \dfrac{1}{3}|2x + 1|$

$2x - 1 = \dfrac{1}{3}(2x + 1)$ $2x - 1 = -\dfrac{1}{3}(2x + 1)$

$6x - 3 = 2x + 1$ $6x - 3 = -(2x + 1)$

$4x = 4$ $6x - 3 = -2x - 1$

$$x = 1 \qquad\qquad 8x = 2$$

$$x = \frac{1}{4}$$

Check:

$$|2(1) - 1| = \frac{1}{3}|2(1) + 1| \qquad\qquad \left|2\left(\frac{1}{4}\right) - 1\right| = \frac{1}{3}\left|2\left(\frac{1}{4}\right) + 1\right|$$

$$1 = \frac{1}{3}|3| \qquad\qquad\qquad \left|-\frac{1}{2}\right| = \frac{1}{3}\left|\frac{3}{2}\right|$$

$$\frac{1}{2} = \frac{1}{2}$$

Both solutions check, so the solution set is $\{\frac{1}{4}, 1\}$.

(Other types of equations)

57. $x^2 - x - 6 \le 0$

$(x - 3)(x + 2) \le 0$

The split points are 3 and –2. Draw a number line with the indicated split points.

The solution is [–2, 3], which can also be written as $-2 \le x \le 3$.

(Other types of inequalities)

58. $2x^2 > 5x + 3$

$2x^2 - 5x - 3 > 0$

$(2x + 1)(x - 3) > 0$

The split points are –1/2 and 3. Draw a number line with the indicated split points.

The solution is $\left(-\infty, -\frac{1}{2}\right) \cup (3, \infty)$ which can also be written as $x < -\frac{1}{2}$ or $x > 3$.

(Other types of inequalities)

59. $(x+1)(x-2)(x+4) \geq 0$

The split points are –1, 2, and –4. Draw a number line with the indicated split points.

The solution is $[-4, -1] \cup [2, \infty)$ which can also be written as $-4 \leq x \leq -1$ or $x \geq 2$.

(Other types of inequalities)

60. $x(x-2)^2(x+1) < 0$

The split points are 0, 2, and –1.

The solution is $(-1, 0)$ which can also be written as $-1 < x < 0$.   **(Other types of inequalities)**

61. $\dfrac{4}{x-1} \leq 2$

$\dfrac{4}{x-1} - 2 \leq 0$ Get 0 on the right side of the inequality.

$\dfrac{4}{x-1} - \dfrac{2(x-1)}{x-1} \leq 0$ Get a common denominator on the left side.

$\dfrac{4-2x+2}{x-1} \leq 0$

$\dfrac{-2x+6}{x-1} \leq 0$

The split points are 3 and 1.

The solution is $(-\infty, 1) \cup [3, \infty]$, which can also be written as $x < 1$ or $x \geq 3$.

Note that $x = 1$ is *not* included in the solution because it makes the denominator equal 0.

(Other types of inequalities)

62.
$$\frac{2}{x+5} > \frac{1}{2x+3}$$

$$\frac{2}{x+5} - \frac{1}{2x+3} > 0 \qquad\qquad \text{Isolate 0.}$$

$$\frac{2(2x+3) - (x+5)}{(x+5)(2x+3)} > 0 \qquad\qquad \text{Combine fractions.}$$

$$\frac{4x+6-x-5}{(x+5)(2x+3)} > 0 \qquad\qquad \text{Multiply.}$$

$$\frac{3x+1}{(x+5)(2x+3)} > 0 \qquad\qquad \text{Combine similar terms.}$$

Find the critical points:

$$3x + 1 = 0 \qquad x + 5 = 0 \qquad 2x + 3 = 0$$

$$x = -\frac{1}{3} \qquad x = -5 \qquad x = -\frac{3}{2}$$

Draw a number line with the indicated critical points.

The solution is $\left(-5, -\frac{3}{2}\right) \cup \left(-\frac{1}{3}, \infty\right)$.

(Other types of inequalities)

Grade Yourself

Circle the question numbers that you had incorrect. Then indicate the number of questions you missed. If you answered more than three questions incorrectly, you will have to focus on that topic. If a topic has fewer than three questions and you had at least one wrong, we suggest you study that topic. Read your textbook or a review book or ask your teacher for help.

Subject: Equations and Inequalities

Topic	Question Numbers	Number Incorrect
Linear equations	1, 2, 3, 4, 5, 6, 7, 8, 9, 10, 11, 12	
Modeling with linear equations	13, 14, 15, 16, 17, 18, 19, 20	
Linear inequalities	21, 22, 23, 24, 25, 26, 27, 28, 29, 30, 31, 32	
Quadratic equations	33, 34, 35, 36, 37, 38, 39, 40, 41, 42, 43	
Other types of equations	44, 45, 46, 47, 48, 49, 50, 51, 52, 53, 54, 55, 56	
Other types of inequalities	57, 58, 59, 60, 61, 62	

Graphs and Functions

Brief Yourself

This chapter includes questions about lines, linear inequalities, circles, and functions.

The formulas needed for working with lines are summarized below:

$m = \dfrac{y_2 - y_1}{x_2 - x_1}$ Slope of the line through (x_1, y_1) and (x_2, y_2).

$y = mx + b$ Slope-intercept equation of a line.

$y - y_1 = m(x - x_1)$ Point-slope equation of a line.

$y = q$ Equation of a horizontal line through (p, q).

$x = p$ Equation of a vertical line through (p, q).

$ax + by = c$ Standard form of the equation of a line.

$Ax + By + C = 0$ General form of the equation of a line.

A summary of some facts about slopes of lines will also be needed:

Parallel lines have equal slopes.

Perpendicular lines have slopes that are negative reciprocals of each other.

A line with a positive slope rises from left to right.

A line with a negative slope falls from left to right.

A line with an undefined slope is a vertical line.

A line with a zero slope is a horizontal line.

Circles

The standard form of a circle with center (h, k) and radius r is $(x - h)^2 + (y - k)^2 = r^2$. To graph a circle when the equation is in standard form, put the center at (h, k), take the square root of the number on the right side and go out that distance from the center to draw the circle. If the equation is not in standard form, complete the square on x and y to find the standard form.

Functions

A function is a set of ordered pairs in which no x-coordinate is repeated. There are various ways to identify functions, depending on the type of information you are given.

1. Given a set of ordered pairs, check the x-coordinates. If no x-coordinate is repeated, the set of ordered pairs describes a function.
2. Given a graph, draw vertical lines through the graph. If no vertical line intersects more than one point on the graph, the graph describes a function. (This is called the vertical line test.)
3. Given a mapping diagram, if each element in the first set maps to only one element in the second set, the map represents a function.
4. Given an equation, graph the equation and use the vertical line test to determine whether the equation represents a function.

Functional Notation

Functions can be written by writing a function name, usually f, g, or h, followed in parentheses by the variable that represents the domain elements: $f(x)$, $g(x)$, $h(x)$ read f of x, g of x, and h of x. Note that $f(x)$ does not mean f times x. In this notation, $f(3)$ represents the y-value paired with the x-value 3.

Functions

The domain of a function is the set of all permissible x-values. Given a set of ordered pairs, the domain is the set of x-coordinates.

Given an equation:

1. Often the domain is the set of all real numbers.
2. If there is a variable in the denominator, set the denominator equal to 0 and solve. The domain consists of all real numbers except the value(s) that makes the denominator 0.
3. If there is a variable in a square root (or an even indexed root), set the radicand greater than or equal to 0 and use that solution as the domain.

The Difference Quotient

The difference quotient $\dfrac{f(x+h) - f(x)}{h}$ can be found by:

1. Replacing x with $(x+h)$ in the given function and simplifying.
2. Subtracting $f(x)$ from the expression found in step 1.
3. Dividing the result by h.

Given the graph of a function, the domain is found by observing the x-values contained in the graph. The range is the set of y-values on the graph. When describing the region in which a graph is increasing, decreasing, or constant, give answers in terms of x-values, even though increasing, decreasing, and constant refer to what is happening to the y-values (increasing means the y-values are increasing, for example). Graphs are also described by types of symmetries. A function is even if $f(-x) = f(x)$. The graph of an even function is symmetric with respect to the y-axis. A function is odd if $f(-x) = -f(x)$. The graph of an odd function is symmetric with respect to the origin.

Test Yourself

Find the slope of the line passing through the given pair of points.

1. (2, 5) and (4,10).

2. (3, 5) and (6, –2).

3. (2, 6) and (2, –4).

4. (–3, 4) and (2, 4)

Graph each line, using the slope and y-intercept.

5. $y = \frac{2}{3}x - 1$

6. $y = \frac{1}{3}x + 2$

7. $3x + y = 4$

8. Write an equation of the line with slope $\frac{2}{3}$ and y-intercept 2.

9. Write an equation of the line with $m = -\frac{1}{2}$ that passes through the point (–1, 3). Write your answer in slope-intercept form.

10. Write an equation of the line containing the points (–3, 4) and (2, –1). Write your answer in slope-intercept form.

11. Write an equation of the horizontal line through (2, –3).

12. Write an equation of the vertical line through (1, 6).

13. Write an equation of the line through (–2, –3) and parallel to $2x + 3y = 6$. Write the answer in general form.

14. Write an equation of the line through (1, –4) and perpendicular to $y = 2x + 4$. Write the answer in general form.

15. Write an equation of the line through (2, 5) and parallel to $x = 4$.

16. Write an equation of the line through (–4, –2) and perpendicular to $x = 2$.

17. Find an equation for the circle with center at (2, 3) and radius 4.

18. Find an equation for the circle with center at the origin and radius 2.

Find the center and radius of each circle.

19. $(x - 2)^2 + (y - 1)^2 = 25$

20. $(x + 3)^2 + (y - 4)^2 = 49$

21. $x^2 + (y - 2)^2 = 16$

22. $x^2 + y^2 = 9$

23. $(x - 1)^2 + (y + 2)^2 = 0$

24. $x^2 + (y - 3)^2 = -4$

Graph each equation.

25. $x^2 + y^2 - 4x + 2y - 4 = 0$

26. $x^2 + y^2 + 6y + 5 = 0$

Identify which of the following are functions.

27. {(2, 5), (1, 3), (2, –5)}

28. {(0, 1), (2, 1), (3, 4)}

29.

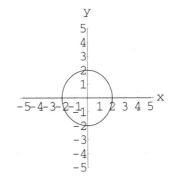

30.

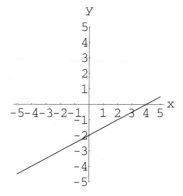

31.

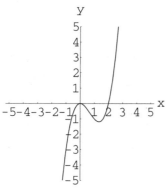

32.

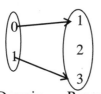

Domain   Range

33.

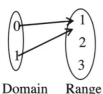

Domain   Range

34.

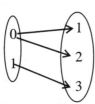

Domain   Range

35. $y = \dfrac{1}{2}x - 3$

36. $y = \sqrt{x - 4}$

Find the domain.

37. $\{(2, 4), (1, 3), (3, 5)\}$

38. $y = 4x - 1$

39. $y = \dfrac{6}{3x - 1}$

40. $y = \sqrt{x + 2}$

41. $y = \dfrac{\sqrt{x + 4}}{x}$

42. Find $g(-2)$ if $g(x) = x^2 - 2x + 1$.

43. Find $f(a)$ if $f(x) = 4x - 6$.

44. Find $f(a + h)$ if $f(x) = x^2 + 4x$.

45. Find $\dfrac{f(x + h) - f(x)}{h}$ and simplify for

$f(x) = x^2 + 4x$.

46. Find $\dfrac{f(x + h) - f(x)}{h}$ and simplify for

$f(x) = \dfrac{1}{x + 1}$.

Use the graph of the function f to find the following.

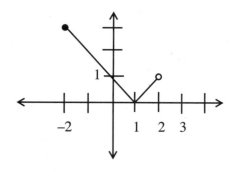

47. The domain and range of f.

48. $f(0)$ and $f(2)$.

49. The increasing, decreasing, or constant behavior of the function.

50. Determine whether $f(x) = 2x^2$ is even, odd, or neither.

51. Determine whether $f(x) = x^3 + 4x$ is even, odd, or neither.

52. Determine whether $f(x) = x^2 + 4x$ is even, odd, or neither.

Check Yourself

1. Let $(2, 5)$ be (x_1, y_1) and $(4, 10)$ be (x_2, y_2). Then substitute into the formula:

$$m = \frac{y_2 - y_1}{x_2 - x_1}$$

$$= \frac{10 - 5}{4 - 2}$$

$$= \frac{5}{2}$$

(Lines)

2. Let $(3, 5) = (x_1, y_1)$ and $(6, -2) = (x_2, y_2)$.

$$m = \frac{-2 - 5}{6 - 3} = -\frac{7}{3}$$

(Lines)

3. Let $(2, 6) = (x_1, y_1)$ and $(2, -4) = (x_2, y_2)$.

$m = \dfrac{-4 - 6}{2 - 2} = -\dfrac{10}{0}$ which is undefined (division by 0). Note that the given points are on a vertical line (they have the same x-coordinate). The slope of any vertical line is undefined.

(Lines)

4. Let $(-3, 4) = (x_1, y_1)$ and $(2, 4) = (x_2, y_2)$.

$m = \dfrac{4 - 4}{2 - (-3)} = \dfrac{0}{5} = 0$. Note that the given points are on a horizontal line (they have the same y-coordinate). The slope of any horizontal line is 0.

(Lines)

5. $y = \frac{2}{3}x - 1$. The y-intercept is $(0, -1)$. The slope is $\frac{2}{3}$ so rise 2 units and run 3 units from $(0, -1)$ to get the

point $(3, 1)$.

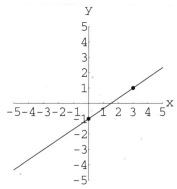

(**Lines**)

6. $y = -\frac{1}{3}x + 2$. The y-intercept is $(0, 2)$. The slope is $-\frac{1}{3}$ so fall 1 unit and run 3 units from $(0, 2)$ to get the

point $(3, 1)$.

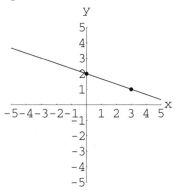

(**Lines**)

7. $3x + y = 4$. Solve the equation for y to get the slope-intercept equation of a line: $y = -3x + 4$. The y-

intercept is 4 and the slope is $-3 = -\frac{3}{1}$.

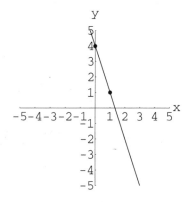

(**Lines**)

8. Use the slope-intercept equation of a line $y = mx + b$: An equation of the line with slope $\frac{2}{3}$ and y-inter-cept 2 is $y = \frac{2}{3}x + 2$.

   (Lines)

9. Since you are given a point and the slope, use the point-slope equation of a line $y - y_1 = m(x - x_1)$:

   $$y - 3 = -\frac{1}{2}(x - (-1))$$

   To write the answer in standard form, the x and y term must be on the left side of the equation:

   $$y - 3 = -\frac{1}{2}(x + 1)$$

   $$2(y - 3) = 2\left(-\frac{1}{2}(x + 1)\right) \qquad \text{Multiply both sides of the equation by 2.}$$

   $$2y - 6 = -x - 1$$

   $$x + 2y = 5 \qquad \text{The answer is now in the standard form of the equation of a line } ax + by = c.$$

   (Lines)

10. When you are given two points, first find the slope using $m = \dfrac{y_2 - y_1}{x_2 - x_1}$:

    $$m = \frac{-1 - 4}{2 - (-3)} = \frac{-5}{5} = -1. \text{ Now use the point-slope equation of a line.}$$

    $$y - 4 = -1(x - (-3))$$

    To write the answer in the slope-intercept form of the equation of a line isolate y:

    $$y - 4 = -x - 3$$

    $$y = -x + 1$$

    (Lines)

11 $y = -3$

    (Lines)

12. $x = 1$

    (Lines)

13. The required line must have the same slope as $2x + 3y = 6$ since parallel lines have the same slopes. Write $2x + 3y = 6$ in slope-intercept form to find its slope:

    $$3y = -2x + 6$$

$$y = -\frac{2}{3}x + 2$$

Now use the point-slope equation of a line with point $(-2, -3)$ and slope $-\frac{2}{3}$:

$$y - (-3) = -\frac{2}{3}(x - (-2))$$

To write the answer in standard form, clear fractions and get the x and y term on the left side of the equation:

$$3(y + 3) = 3\left(-\frac{2}{3}(x + 2)\right)$$

$$3y + 9 = -2x - 4$$

$$2x + 3y = -13$$

(Lines)

14. To be perpendicular to $y = 2x + 4$, a line must have slope $-\frac{1}{2}$. Now use the point-slope equation of a line:

$$y - (-4) = -\frac{1}{2}(x - 1)$$

To write the answer in slope-intercept equation of a line isolate y:

$$y + 4 = -\frac{1}{2}x + \frac{1}{2}$$

$$y = -\frac{1}{2}x - \frac{7}{2}$$

(Lines)

15. Since $x = 4$ is a vertical line, any line parallel to it must also be a vertical line. An equation of the vertical line through $(2, 5)$ is $x = 2$.

(Lines)

16. Since $x = 2$ is a vertical line, any line perpendicular to it must be a horizontal line. An equation of the horizontal line through $(-4, -2)$ is $y = -2$.

(Lines)

17. $(h, k) = (2, 3)$ $r = 4$ Identify h, k, and r.

    $(x - h)^2 + (y - k)^2 = r^2$ Write the equation of a circle.

    $(x - 2)^2 + (y - 3)^2 = (4)^2$ Substitute for h, k, and r.

    $(x - 2)^2 + (y - 3)^2 = 16$

(Circles)

18. $(h, k) = (0, 0)$ $r = 2$ Center at the origin means $(h, k) = (0, 0)$.

       $(x - h)^2 + (y - k)^2 = r^2$ Write the equation of a circle.

       $(x - 0)^2 + (y - 0)^2 = (2)^2$ Substitute for h, k, and r.

       $x^2 + y^2 = 4$ Simplify.

       (Circles)

19. $(x - 2)^2 + (y - 1)^2 = 25$

       $(x - 2)^2 + (y - 1)^2 = (5)^2$ Write in the form $(x - h)^2 + (y - k)^2 = r^2$.

       $h = 2, k = 1, r = 5$ Identify h, k, and r.

       The center is at (2, 1) and the radius is 5.

       (Circles)

20. $(x + 3)^2 + (y - 4)^2 = 49$

       $(x - (-3))^2 + (y - 4)^2 = (7)^2$ Write in the form $(x - h)^2 + (y - k)^2 = r^2$.

       $h = -3, k = 4, r = 7$ Identify h, k, and r.

       The center is at (–3, 4) and the radius is 7.

       (Circles)

21. $x^2 + (y - 2)^2 = 16$

       $(x - 0)^2 + (y - 2)^2 = (4)^2$ Write in the form $(x - h)^2 + (y - k)^2 = r^2$.

       $h = 0, k = 2, r = 4$ Identify h, k, and r.

       The center is at (0, 2) and the radius is 4.

       (Circles)

22. $x^2 + y^2 = 9$

       $(x - 0)^2 + (y - 0)^2 = 3^2$ Write in the form $(x - h)^2 + (y - k)^2 = r^2$.

       $h = 0, k = 0, r = 3$ Identify h, k, and r.

       The center is at (0, 0) and the radius is 3.

       (Circles)

23. $(x - 1)^2 + (y + 2)^2 = 0$

       Note that the radius is $\sqrt{0} = 0$. This is called a degenerate circle and consists of one point, (1, –2).

       (Circles)

24. $x^2 + (y-3)^2 = -4$

Note that the radius is $\sqrt{-4}$ which is an imaginary number. There are no points that satisfy this equation, therefore there is no center or radius.

(Circles)

25. $x^2 + y^2 - 4x + 2y - 4 = 0$

First rewrite the equation in the form $(x-h)^2 + (y-k)^2 = r^2$ to find the center and radius.

$x^2 - 4x + y^2 + 2y = 4$ Complete the square on x and on y.

$x^2 - 4x + \mathbf{4} + y^2 + 2y + \mathbf{1} = 4 + \mathbf{4} + \mathbf{1}$ $(\frac{1}{2} \cdot (-4))^2 = 4$ $(\frac{1}{2} \cdot 2)^2 = 1$

$(x-2)^2 + (y+1)^2 = 9$

$(x-2)^2 + (y-(-1))^2 = 3^2$ Write in the form $(x-h)^2 + (y-k)^2 = r^2$.

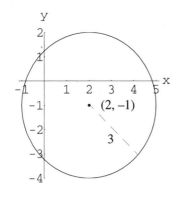

The center is at $(2, -1)$ and the radius is 3.

(Circles)

26. $x^2 + y^2 + 6y + 5 = 0$

     $x^2 + y^2 + 6y = -5$ We only need to complete the square on y.

     $x^2 + (y^2 + 6y + 9) = -5 + 9$ $(\frac{1}{2} \cdot 6)^2 = 9$

        $x^2 + (y+3)^2 = 4$

$(x+0)^2 + (y-(-3))^2 = (2)^2$

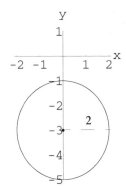

The center is at (0, −3) and the radius is 2.

(Circles)

27.   Not a function. Note the repeated *x*-value of 2.

     (Functions)

28.   This is a function. There are no repeated *x*-values.

     (Functions)

29.   Not a function. The graph does not pass the vertical line test.

     (Functions)

30.   This is a function. The graph passes the vertical line test.

     (Functions)

31.   This is a function. The graph passes the vertical line test.

     (Functions)

32.   This is a function. No element in the domain is used more than once.

     (Functions)

33.   This is a function. The set of ordered pairs is {(0, 1), (1, 3)}.

     (Functions)

34.   Not a function. The set of ordered pairs is {(0, 1), (0, 2), (1, 3)}. The element 0 is used twice.

     (Functions)

35.   Draw the graph. Note that it passes the vertical line test. Yes, this is a function.

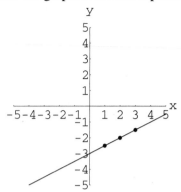

(Functions)

36.

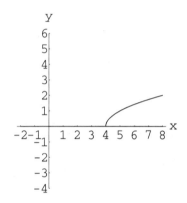

Draw the graph. Note that it passes the vertical line test. Yes, this is a function. Also note that for each $x \geq 4$, only one y-value is specified.

(Functions)

37.   The domain is {2, 1, 3}.

(Functions)

38.   The domain is all real numbers written $(-\infty, \infty)$.

(Functions)

39. $3x - 1 = 0$ Set the denominator equal to 0 and solve.

       $3x = 1$

       $x = \dfrac{1}{3}$

       The domain is all real numbers except $\dfrac{1}{3}$.

(Functions)

40. $x + 2 \geq 0$ Set the radicand greather than or equal to 0 and solve.

$x \geq -2$

The domain is $x \geq -2$.

(Functions)

41. $x + 4 \geq 0$ *and* $x \neq 0$ (since this would make the denominator undefined)

$x \geq -4$

The domain is all real numbers greater than or equal to -4 except 0.

(Functions)

42. $g(-2) = (-2)^2 - 2(-2) + 1$

$= 4 + 4 + 1$

$= 9$

(Functions)

43. $f(a) = 4(a) - 6 = 4a - 6$

(Functions)

44. $f(a + h) = (a + h)^2 + 4(a + h)$ Replace each occurrence of x with $(a + h)$.

$= a^2 + 2ah + h^2 + 4a + 4h$

(Functions)

45. $f(x) = x^2 + 4x$

$f(x + h) = (x + h)^2 + 4(x + h)$ Find $f(x + h)$.

$= x^2 + 2xh + h^2 + 4x + 4h$ Simplify.

$\dfrac{x^2 + 2xh + h^2 + 4x + 4h - (x^2 + 4x)}{h}$ $\dfrac{f(x + h) - f(x)}{h}$

$= \dfrac{x^2 + 2xh + h^2 + 4x + 4h - x^2 - 4x}{h}$

$= \dfrac{2xh + h^2 + 4h}{h}$ Combine like terms.

$= \dfrac{h(2x + h + 4)}{h}$ Factor out h.

$= 2x + h + 4$ Cancel.

(Functions)

46. $f(x) = \dfrac{1}{x+1}$

$f(x+h) = \dfrac{1}{x+h+1}$ Find $f(x+h)$.

$\dfrac{\dfrac{1}{x+h+1} - \dfrac{1}{x+1}}{h}$ $\dfrac{f(x+h) - f(x)}{h}$

$= \dfrac{(x+1)(x+h+1) \cdot \dfrac{1}{x+h+1} - (x+1)(x+h+1) \cdot \dfrac{1}{x+1}}{(x+1)(x+h+1)h}$

$= \dfrac{x+1 - (x+h+1)}{(x+1)(x+h+1)h}$ Cancel.

$= \dfrac{x+1 - x - h - 1}{(x+1)(x+h+1)h}$ Distribute.

$= \dfrac{-h}{(x+1)(x+h+1)h}$ Combine like terms.

$= \dfrac{-1}{(x+1)(x+h+1)}$ Reduce.

(Functions)

47.   The domain is [–2, 2). Note that the open dot means the point (2, 1) is not included in the graph. The range is [0, 3].

(Graphs of functions)

48. $f(0) = 1$. That is, when $x = 0, y = 1$. $f(2)$ = does not exist. Since (2, 1) is an open dot, it is *not* part of the graph.

(Graphs of functions)

49. f is decreasing on (–2, 1) and increasing on (1, 2). Note that increasing and decreasing intervals do not include end points.

(Graphs of functions)

50. $f(x) = 2x^2$

$f(-x) = 2(-x)^2 = 2x^2$

Since $f(-x) = f(x)$, the function is even.

(Graphs of functions)

51. $f(x) = x^3 + 4x$

$f(-x) = (-x)^3 + 4(-x) = -x^3 - 4x$

Since $f(-x) = -f(x)$, the function is odd.

(Graphs of functions)

52. $f(x) = x^2 + 4x$

$f(-x) = (-x)^2 + 4(-x) = x^2 - 4x$

Since $f(-x) \neq f(x)$ or $-f(-x)$, the function is neither odd nor even.

(Graphs of functions)

Grade Yourself

Circle the question numbers that you had incorrect. Then indicate the number of questions you missed. If you answered more than three questions incorrectly, you will have to focus on that topic. If a topic has fewer than three questions and you had at least one wrong, we suggest you study that topic. Read your textbook or a review book or ask your teacher for help.

Subject: Graphs and Functions

Topic	Question Numbers	Number Incorrect
Lines	1, 2, 3, 4, 5, 6, 7, 8, 9, 10, 11, 12, 13, 14, 15, 16	
Circles	17, 18, 19, 20, 21, 22, 23, 24, 25, 26	
Functions	27, 28, 29, 30, 31, 32, 33, 34, 35, 36, 37, 38, 39, 40, 41, 42, 43, 44, 45, 46	
Graphs of functions	47, 48, 49, 50, 51, 52	

Functions

4

Brief Yourself

This chapter contains additional questions about functions, including translations and combinations of functions, the algebra of functions, inverse functions, the graphs of polynomial and rational functions, and variation.

Even if you use a graphing calculator, you should recognize and be able to quickly sketch the graphs of the following functions:

Constant function

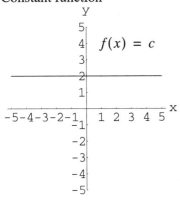

$f(x) = c$

Identity function

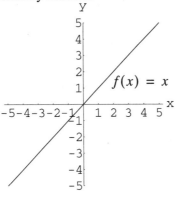

$f(x) = x$

Absolute value function

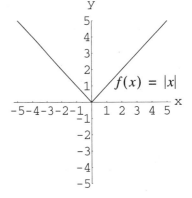

$f(x) = |x|$

Square root function

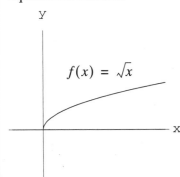

$f(x) = \sqrt{x}$

Squaring function

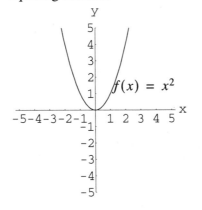

$f(x) = x^2$

Cubing function

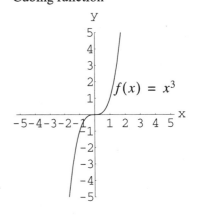

$f(x) = x^3$

These basic shapes can be shifted horizontally and vertically as indicated in the following chart. For $a > 0$, and $f(x)$ a function:

$f(x) + a$ is shifted up a units.

$f(x) - a$ is shifted down a units.

$f(x + a)$ is shifted left a units.

$f(x - a)$ is shifted right a units.

$-f(x)$ is reflected in the x-axis.

$f(-x)$ is reflected in the y-axis.

Graphs can also be stretched vertically with $cf(x)$ where $c > 1$, and shrunk vertically where $0 < c < 1$.

Functions can be combined by the operations of addition, subtraction, multiplication, and division, as well as composition. Composition of functions indicates the order in which the functions are evaluated; for example, $(fog)(x)$ means first evaluate $g(x)$ and then evaluate f at that answer.

Questions about inverse functions include:

1. Using the horizontal line test to determine when an inverse function exists (any horizontal line passes through the graph of a one-to-one function at most once).
2. Finding an inverse function by:

    reversing the coordinates in a set of ordered pairs.

    interchanging x and y and solving for the new y if given an equation

    reflecting a graph across the line $y = x$.
3. Verifying that two functions f and g are inverses using $f(g(x)) = x = g(f(x))$.

To graph quadratic functions that are in standard form $y = a(x - h)^2 + k$, use the shifting and stretching concepts to change $y = x^2$ to the desired shape and location. These changes are controlled by h, k, and a as follows:

shift left h units if $h > 0$.

shift right h units if $h < 0$.

shift up k units if $k > 0$.

shift down k units if $k < 0$.

open up if $a > 0$.

open down if $a < 0$.

stretch vertically if $a > 1$.

shrink vertically if $0 < a < 1$.

If a quadratic function is not in standard form, complete the square to write it in $y = a(x - h)^2 + k$ form. Any x-intercept(s) can be found by setting $y = 0$, and the y-intercept can be found by setting $x = 0$.

When graphing polynomial functions, be sure your graphs are smooth curves (no sharp points) and continuous (no breaks). The shifting and stretching techniques can be applied to the basic graph of $f(x) = x^n$, where the graph resembles a parabola when n is even and resembles a cubic when n is odd. The leading coefficient determines the right and left hand behavior of the graph as indicated below for $f(x) = ax^n + \ldots$

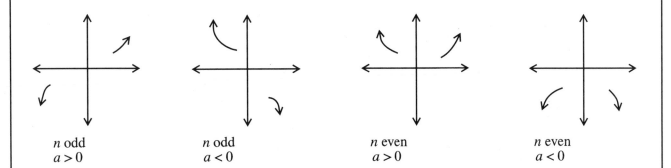

n odd	n odd	n even	n even
$a > 0$	$a < 0$	$a > 0$	$a < 0$

The x-intercepts of a polynomial function are referred to by various terms. You should be able to recognize what each term refers to:

If $(a, 0)$ is an *x-intercept* of a polynomial function f then

> $x = a$ is a *zero* of the function f
>
> $x = a$ is a *solution* of $f(x) = 0$
>
> $(x - a)$ is a *factor* of $f(x)$.

To graph rational functions:

1. Graph dotted lines to show vertical asymptotes (these occur when the denominator is equal to 0).
2. Graph a dotted line at the horizontal asymptote (rules for horizontal asymptotes follow).
3. Find and plot x- and y-intercepts.
4. Use a table of values to plot at least one point on each side of the vertical asymptotes.
5. Connect points with a smooth curve.

Never connect across a vertical asymptote.

To find horizontal asymptotes, for $f(x) = \dfrac{ax^m + \ldots}{bx^n + \ldots}$

1. If $m > n$, there is no horizontal asymptote.

2. If $m = n$, $y = \dfrac{a}{b}$ is the horizontal asymptote.

3. If $m < n$, $y = 0$ is the horizontal asymptote.

Variation

To translate word problems involving variation, you will need three basic formulas.

Direct variation: y varies directly as x or y is proportional to x: use $y = kx$.

Indirect variation: y varies inversely as x or y varies indirectly as x: use $y = \dfrac{k}{x}$.

Joint variation: y varies jointly as x and z: use $y = kxz$.

Test Yourself

Sketch the graph of each function.

1. $f(x) = x^3 + 1$

2. $f(x) = |x + 1|$

3. $g(x) = \sqrt{x + 2} - 1$

4. $h(x) = -(x - 2)^2 + 3$

5. Describe how the graph of each function would differ from the graph of $f(x) = \sqrt[4]{x}$. Do not sketch the graphs.

    (a) $f(x) = \sqrt[4]{x} - 1$

    (b) $f(x) = \dfrac{1}{2}\sqrt[4]{x}$

    (c) $f(x) = \sqrt[4]{x - 1}$

6. Sketch the graph of $g(x) = \begin{cases} |x - 1| & \text{if } x < 3 \\ |x - 3| & \text{if } x \geq 3 \end{cases}$.

7. Let $f(x) = 4x - 1$ and $g(x) = x^2 + 2x$. Find $(f + g)(x)$, $(f - g)(x)$, $(fg)(x)$, and $\left(\dfrac{f}{g}\right)(x)$.

8. Let $f(x) = x^2 + 4$ and $g(x) = \sqrt{4 - x}$. Find $(f + g)(3)$, $(f - g)(2)$, $(fg)(0)$, and $\left(\dfrac{f}{g}\right)(3)$.

9. Let $f(x) = x^2$ and $g(x) = 2x + 1$. Find $(fog)(x)$ and $(gof)(x)$.

10. Let $f(x) = \dfrac{1}{x}$ and $g(x) = \dfrac{1}{x + 2}$. Find $(fog)(x)$ and $(gof)(x)$.

11. Is the graph below the graph of a one-to-one function?

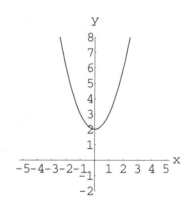

12. Is the graph below the graph of a one-to-one function?

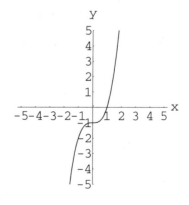

13. Is the graph below the graph of a one-to-one function?

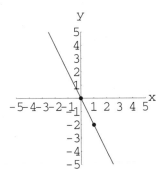

Find the inverse.

14. $\{(2, 5), (1, 3), (4, -6)\}$

15. $y = 3x + 2$

16. $f(x) = \frac{2}{3}x - 4$

17. $f(x) = x^3 + 1$

18.

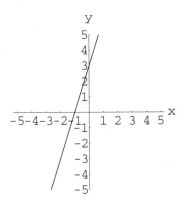

19.

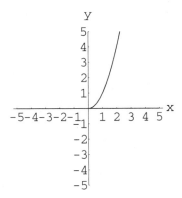

20. Determine whether $h(x) = \sqrt{x - 2}$ has an inverse. If so, find it.

Describe how the graph of g differs from the graph of $f(x) = x^2$.

21. $g(x) = -\frac{1}{2}x^2$

22. $g(x) = 2(x + 1)^2 - 3$

Find the vertex and intercepts of each parabola and sketch the graph.

23. $f(x) = -(x - 2)^2 + 4$

24. $f(x) = x^2 + 6x + 8$

25. $h(x) = 2x^2 - 4x - 6$

26. Find the quadratic function that has vertex $(-1, 3)$ and passes through the point $(0, 1)$.

27. The profit for a company is given by $P(x) = 20x - 0.02x^2 - 320$ where x is the number of units sold. What sales level will yield a maximum profit?

28. Describe how the graph of g differs from the graph of $f(x) = x^4$ if $g(x) = (x + 1)^4$.

29. Describe how the graph of g differs from the graph of $f(x) = x^5$ if $g(x) = (x - 2)^5 + 1$.

Determine the right-hand and left-hand behavior of the graph of the polynomial function.

30. $f(x) = 4x^3 + 5x^2 - 2x$

31. $f(x) = -6x^4 + 2x^2 + 3x - 1$

32. $f(x) = x^6 + 4x^3 + x^2 - 8$

33. $f(x) = 2 - x^5$

34. Find all the real zeros of $h(t) = t^3 - 4t^2 + 4t$.

35. Find a polynomial function that has zeros 2, -1, 1, and 0.

Sketch the graph of each function.

36. $f(x) = x(x - 2)(x + 1)$

37. $f(x) = 2 - x^5$

38. $f(x) = x^2(x + 4)$

39. $f(x) = (x + 3)^6 - 2$

40. Find the horizontal and vertical asymptotes for
$$f(x) = \frac{x - 2}{x^2 + x - 12}.$$

41. Find the horizontal and vertical asymptotes for
$$f(x) = \frac{x^2}{2x^2 - 5x - 3}.$$

42. Find the horizontal and vertical asymptotes for

$$f(x) = \frac{3x^2 + 1}{x}$$

Sketch the graph of each rational function. Label any asymptotes and intercepts.

43. $f(x) = \dfrac{2}{x - 4}$

44. $g(x) = \dfrac{x - 3}{x^2 - 4}$

45. If x is 20 when y is 90 and y varies directly as x, find y when x is 40.

46. Hooke's Law states that the force needed to stretch a spring is proportional to the distance it is stretched. If a force of 5 pounds stretches a spring a distance of 2 cm, find the force needed to stretch the same spring 12 cm.

47. Boyle's Law states that the pressure of a compressed gas is inversely proportional to the volume of the gas. If there is a pressure of 28 pounds per square inch when the volume of gas is 50 cubic inches, find the pressure when the gas has a volume of 200 cubic inches.

48. If r varies jointly as s and t^3, and r is 36 when s is 4 and t is 2, find r when s is 3 and t is 4.

✔ Check Yourself

1. This is the graph of $f(x) = x^3$ shifted up 1 unit: $f(x) = x^3 + 1$

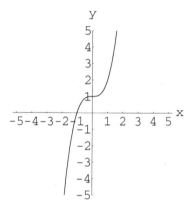

(Translations and combinations)

2. This is the graph of $f(x) = |x|$ shifted to the left 1 unit:

$$f(x) = |x + 1|$$

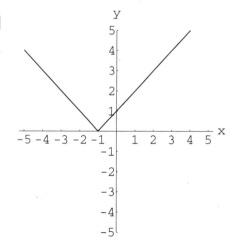

(Translations and combinations)

3.   This is the graph of $f(x) = \sqrt{x}$ shifted to the left 2 units and down 1 unit:

$g(x) = \sqrt{x+2} - 1$

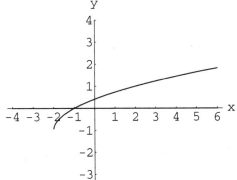

(Translations and combinations)

4.   This is the graph of $f(x) = x^2$ reflected in the *x*-axis, shifted to the right 2 units, and shifted up 3 units:

$h(x) = -(x-2)^2 + 3$

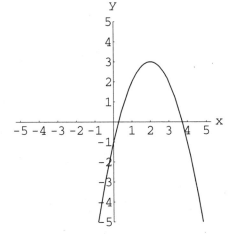

(Translations and combinations)

5. (a) $f(x) = \sqrt[4]{x} - 1$ is shifted down 1 unit from the graph of $f(x) = \sqrt[4]{x}$.

   (b) $f(x) = \frac{1}{2}\sqrt[4]{x}$ is shrunk vertically from the graph of $f(x) = \sqrt[4]{x}$.

   (c) $f(x) = \sqrt[4]{x-1}$ is shifted to the right 1 unit from the graph of $f(x) = \sqrt[4]{x}$.

   (Translations and combinations)

6.

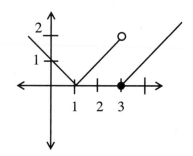

(Translations and combinations)

7. $f(x) = 4x - 1$ and $g(x) = x^2 + 2x$

$(f + g)(x) = (4x - 1) + (x^2 + 2x) = x^2 + 6x - 1$

$(f - g)(x) = (4x - 1) - (x^2 + 2x) = -x^2 + 2x - 1$

$(fg)(x) = (4x - 1)(x^2 + 2x) = 4x^3 + 8x^2 - x^2 - 2x = 4x^3 + 7x^2 - 2x$

$\left(\dfrac{f}{g}\right)(x) = \dfrac{4x - 1}{x^2 + 2x}, x \neq 0, -2$

(Algebra of functions)

8. $f(x) = x^2 + 4$ and $g(x) = \sqrt{4 - x}$

$(f + g)(x) = x^2 + 4 + \sqrt{4 - x}$ so $(f + g)(3) = 3^2 + 4 + \sqrt{4 - 3} = 14$

$(f - g)(x) = x^2 + 4 - \sqrt{4 - x}$ so $(f - g)(2) = 2^2 + 4 - \sqrt{4 - 2} = 8 - \sqrt{2}$

$(fg)(x) = (x^2 + 4)\sqrt{4 - x}$ so $(fg)(0) = (0^2 + 4)\sqrt{4 - 0} = 8$

$\left(\dfrac{f}{g}\right)(x) = \dfrac{x^2 + 4}{\sqrt{4 - x}}$ so $\left(\dfrac{f}{g}\right)(3) = \dfrac{3^2 + 4}{\sqrt{4 - 3}} = \dfrac{13}{1} = 13$

(Algebra of functions)

9. $f(x) = x^2$ and $g(x) = 2x + 1$

$(fog)(x) = f(g(x)) = f(2x + 1) = (2x + 1)^2 = 4x^2 + 4x + 1$

$(gof)(x) = g(f(x)) = g(x^2) = 2(x^2) + 1 = 2x^2 + 1$

(Algebra of functions)

10. $f(x) = \dfrac{1}{x}$ and $g(x) = \dfrac{1}{x + 2}$

$$(f \circ g)(x) = f\left(\frac{1}{x+2}\right) = \frac{1}{\dfrac{1}{x+2}} = x + 2$$

$$(g \circ f)(x) = g\left(\frac{1}{x}\right) = \frac{1}{\dfrac{1}{x} + 2} = \frac{x \cdot 1}{x \cdot \dfrac{1}{x} + x \cdot 2} = \frac{x}{1 + 2x}$$

(Algebra of functions)

11. No. Draw several horizontal lines and observe that some of them intersect at more than one point on the graph.

    (Inverse functions)

12. Yes. The graph passes the horizontal line test.

    (Inverse functions)

13. Yes.

    (Inverse functions)

14. $\{(5, 2), (3, 1), (-6, 4)\}$ Interchange the *x*- and *y*-coordinates.

    (Inverse functions)

15. $\quad y = 3x + 2$

    $\quad x = 3y + 2$ Interchange *x* and *y*.

    $\quad \dfrac{x - 2}{3} = y$ Solve for *y*.

    $\quad f^{-1}(x) = \dfrac{x - 2}{3}$ Use the notation for the inverse function.

    (Inverse functions)

16. $f(x) = \dfrac{2}{3}x - 4$ can be written $y = \dfrac{2}{3}x - 4$.

    $\quad x = \dfrac{2}{3}y - 4$ Interchange *x* and *y*.

    $\quad x + 4 = \dfrac{2}{3}y$ Solve for *y*.

    $\quad \dfrac{3}{2}(x + 4) = y$

    $\quad f^{-1}(x) = \dfrac{3}{2}(x + 4)$ Use the notation for the inverse function.

    (Inverse functions)

17. $y = x^3 + 1$ Set $y = f(x)$.

         $x = y^3 + 1$ Interchange x and y.

         $x - 1 = y^3$

         $\sqrt[3]{x - 1} = y$ Solve for y.

         $f^{-1}(x) = \sqrt[3]{x - 1}$ Use the notation for the inverse function.

         (Inverse functions)

18.   Reflect the given graph across the line $y = x$.

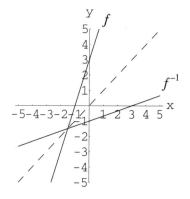

         (Inverse functions)

19.

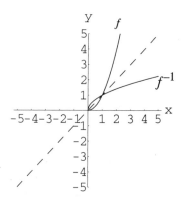

         (Inverse functions)

20.   The graph of $h(x) = \sqrt{x - 2}$ is the same as the graph of $f(x) = \sqrt{x}$ shifted 2 units to the right. The graph passes the horizontal line test, so the inverse exists.

         $y = \sqrt{x - 2}$

         $x = \sqrt{y - 2}$ Interchange x and y.

         $x^2 = \left(\sqrt{y - 2}\right)^2$ Square both sides.

         $x^2 = y - 2$

         $x^2 + 2 = y$ Solve for y.

$$f^{-1}(x) = x^2 + 2$$ Write the answer using inverse notation.

(Inverse functions)

21. The graph of $g(x) = -\dfrac{1}{2}x^2$ is reflected in the x-axis from the graph of $f(x) = x^2$. The graph of $g(x)$ is

    also shrunk vertically from the graph of $f(x) = x^2$.

    (Parabolas)

22. The graph of $g(x) = 2(x+1)^2 - 3$ is shifted 1 unit to the left and down 3 units from the graph of

    $f(x) = x^2$. It is also stretched vertically from the graph of f.

    (Parabolas)

23. $f(x) = -(x-2)^2 + 4$ is reflected in the x-axis, shifted 2 units to the right and up 4 units from the graph of

    $f(x) = x^2$. The vertex is $(2, 4)$.

    Find the intercepts.

    $y = -(0-2)^2 + 4$ Set $x = 0$.

    $y = -(-2)^2 + 4$

    $y = 0$ y-intercept.

    $\quad 0 = -(x-2)^2 + 4$ Set $y = 0$.

    $\quad -4 = -(x-2)^2$

    $\quad\quad 4 = (x-2)^2$

    $\pm\sqrt{4} = (x-2)$ Take the square root of both sides.

    $\quad \pm 2 = x - 2$

    $x - 2 = 2$ or $x - 2 = -2$

    $\quad\quad x = 4$ or $\quad x = 0$ x-intercepts.

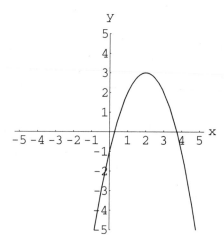

(Parabolas)

24. $f(x) = x^2 + 6x + 8$ Complete the square.

$f(x) + 9 = (x^2 + 6x + 9) + 8$ $\left(\dfrac{1}{2} \cdot 6\right)^2 = 3^2 = 9$ Add 9 to each side.

$f(x) + 9 = (x + 3)^2 + 8$ Factor.

$f(x) = (x + 3)^2 - 1$ Solve for $f(x)$.

The vertex is $(-3, -1)$.

Find the intercepts:

$y = 0^2 + 6(0) + 8$ Set $x = 0$.

$y = 8$ y-intercept.

$0 = x^2 + 6x + 8$ Set $y = 0$.

$0 = (x + 2)(x + 4)$ Factor.

$x = -2$ or $x = -4$ Set each factor equal to 0 and solve.

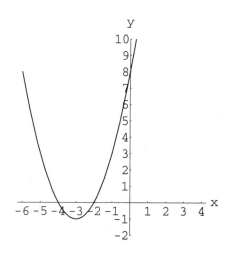

Notice that this is the parabola $y = x^2$ shifted 3 units to the left and shifted down 1 unit.

(Parabolas)

25. Use the process of completing the square:

$$h(x) = 2x^2 - 4x - 6$$

$$h(x) = 2(x^2 - 2x) - 6$$

$$h(x) + 2 = 2(x^2 - 2x + 1) - 6 \qquad \left(\frac{1}{2} \cdot -2\right)^2 = 1$$

$$h(x) + 2 = 2(x - 1)^2 - 6 \qquad \text{Factor.}$$

$$h(x) + 2 = 2(x - 1)^2 - 8 \qquad \text{Solve for } h(x).$$

The vertex is (1, –8).

Find the intercepts:

$$y = 2(0)^2 - 4(0) - 6 \qquad \text{Set } x = 0.$$

$$y = -6 \qquad \qquad y\text{-intercept.}$$

$$0 = 2x^2 - 4x - 6 \qquad \text{Set } y = 0.$$

$$0 = 2(x^2 - 2x - 3) \qquad \text{Factor.}$$

$$0 = 2(x - 3)(x + 1)$$

$$x = 3 \text{ or } x = -1 \qquad x\text{-intercepts.}$$

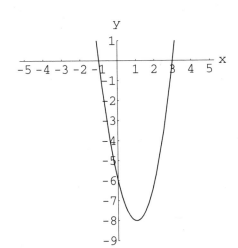

(Parabolas)

26. Use standard form with a vertex of (–1, 3):

$$y = a(x + 1)^2 + 3$$

Since the parabola contains $(0, 1)$, substitute $x = 0$ and $y = 1$:

$$1 = a(0 + 1)^2 + 3$$

$$1 = a + 3 \qquad \text{Solve for } a.$$

$$-2 = a$$

Therefore the equation is $y = -2(x + 1)^2 + 3$.

(Parabolas)

27.   The maximum profit will occur at the vertex of the parabola $y = 20x - 0.02x^2 - 320$. Write the equation in standard form:

$$y = -0.02x^2 + 20x - 320$$

$$y = -0.02(x^2 - 1000x) - 320$$

$$y - 5000 = -0.02(x^2 - 1000x + 250000) - 320$$

$$y - 5000 = -0.02(x - 500)^2 - 320$$

$$y = -0.02(x - 500)^2 + 4680$$

The vertex occurs at $(500, 4680)$ which means when 500 units are sold, the profit is a maximum.

(Parabolas)

28.   The graph of $g(x) = (x + 1)^4$ is shifted 1 unit to the left from the graph of $f(x) = x^4$.

(Graphing polynomials)

29.   The graph of $g(x) = (x - 2)^5 + 1$ is shifted 2 units to the right and up 1 unit from the graph of $f(x) = x^5$.

(Graphing polynomials)

30. $f(x) = 4x^3 + 5x^2 - 2x$. The degree is odd and the leading coefficient is positive. Therefore the graph falls to the left and rises to the right.

(Graphing polynomials)

31. $f(x) = -6x^4 + 2x^2 + 3x - 1$. The degree is even and the leading coefficient is negative. The graph falls to the left and falls to the right.

(Graphing polynomials)

32. $f(x) = x^6 + 4x^3 + x^2 - 8$. The degree is even and the leading coefficient is positive. The graph rises to the left and rises to the right.

(Graphing polynomials)

33. $f(x) = 2 - x^5 = -x^5 + 2$. The degree is odd and the leading coefficient is negative. The graph rises to the left and falls to the right.

     (Graphing polynomials)

34. "Find all the real zeros" is equivalent to "find the *x*-intercepts." Set $h(t) = 0$.

     $0 = t^3 - 4t^2 + 4t$

     $0 = t(t^2 - 4t + 4)$ Factor.

     $0 = t(t - 2)^2$ Factor.

     $t = 0$ $t = 2$ Set each factor equal to 0 and solve.

     (Graphing polynomials)

35. If the zeros are 2, –1, 1, and 0, the polynomial has factors:

     $(x - 2)(x - (-1))(x - 1)(x) = f(x)$

     $f(x) = (x - 2)(x + 1)(x - 1)x$

     $f(x) = (x - 2)(x^2 - 1)x$ Multiply.

     $f(x) = (x^3 - 2x^2 - x + 2)x$

     $f(x) = x^4 - 2x^3 - x^2 + 2x$

     (Graphing polynomials)

36.

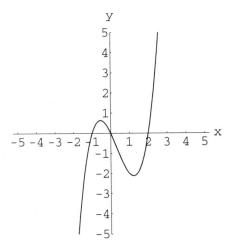

     (Graphing polynomials)

37.

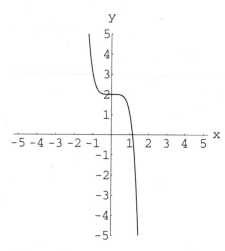

(Graphing polynomials)

38.

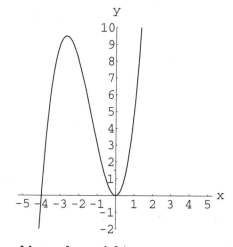

(Graphing polynomials)

39.

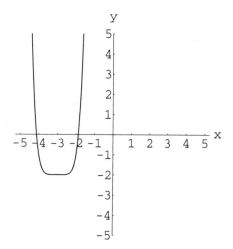

(Graphing polynomials)

40. $f(x) = \dfrac{x - 2}{x^2 + x - 12}$

Vertical asymptotes:

$$x^2 + x - 12 = 0$$
Set the denominator equal to 0.

$$(x + 4)(x - 3) = 0$$
Factor.

$$x = -4 \text{ or } x = 3$$
Solve.

Horizontal asymptotes:

Since the degree of the numerator is less than the degree of the denominator, $y = 0$ is the horizontal asymptote.

(Rational functions)

41. $f(x) = \dfrac{x^2}{2x^2 - 5x - 3}$

Vertical asymptotes:

$$2x^2 - 5x - 3 = 0$$
Set the denominator equal to 0.

$$(2x + 1)(x - 3) = 0$$
Factor.

$$x = -\frac{1}{2} \text{ or } x = 3$$
Solve.

Horizontal asymptote:

Since the degree of the numerator equals the degree of the denominator, $y = \dfrac{1}{2}$ is the horizontal asymptote.

(Rational functions)

42. $f(x) = \dfrac{3x^2 + 1}{x}$

Vertical asymptote:

$$x = 0$$
Set the denominator equal to 0.

Horizontal asymptote:

Since the degree of the numerator is greater than the degree of the denominator, there is no horizontal asymptote.

(Rational functions)

43. $y = \dfrac{2}{x-4}$

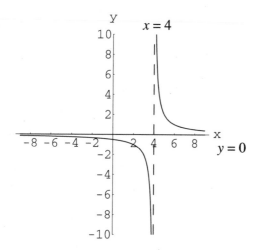

(Rational functions)

44. $y = \dfrac{x-3}{x^2-4}$

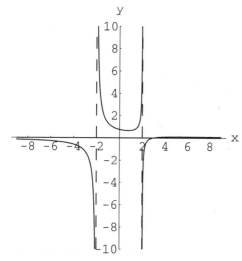

(Rational functions)

45. Since y varies directly as x, use $y = kx$. Use the value of $x = 20$ when $y = 90$ to find the value of k:

$$y = kx$$
$$90 = k(20)$$
$$\frac{90}{20} = k$$
$$\frac{9}{2} = k$$

Now find y when $x = 40$:

$$y = \frac{9}{2}(40)$$

$$= 180$$

(Variation)

46. Let F be the force and x be the distance the spring is stretched. Then $F = kx$. Find the value of k:

$$F = kx$$

$$5 = k(2)$$

$$\frac{5}{2} = k$$

Now find F when $x = 12$:

$$F = \frac{5}{2}(12) = 30 \text{ pounds.}$$

(Variation)

47. Let P = pressure and V = volume. Then Boyle's Law is

$$P = \frac{k}{V}.$$

Find k:

$$28 = \frac{k}{50}$$

$$28(50) = k$$

$$1,400 = k$$

Now use the value of k to find P when $V = 200$:

$$P = \frac{1,400}{200} = 7$$

Therefore, the pressure is 7 pounds per square inch.

(Variation)

48. Use the formula for joint variation to write: $r = kst^3$

Find k when $r = 36$ and $s = 4$ and $t = 2$:

$$36 = k(4)(2)^3$$

$$36 = k(32)$$

$$\frac{36}{32} = k$$

$$\frac{9}{8} = k$$

Now find r when $s = 3$ and $t = 4$:

$$r = \frac{9}{8}(3)(4)^3$$

$$r = \frac{9}{8}(3)(64)$$

$$r = 216$$

(Variation)

Grade Yourself

Circle the question numbers that you had incorrect. Then indicate the number of questions you missed. If you answered more than three questions incorrectly, you will have to focus on that topic. If a topic has fewer than three questions and you had at least one wrong, we suggest you study that topic. Read your textbook or a review book or ask your teacher for help.

Subject: Functions

Topic	Question Numbers	Number Incorrect
Translations and combinations	1, 2, 3, 4, 5, 6	
Algebra of functions	7, 8, 9, 10	
Inverse functions	11, 12, 13, 14, 15, 16, 17, 18, 19, 20	
Parabolas	21, 22, 23, 24, 25, 26, 27	
Graphing polynomials	28, 29, 30, 31, 32, 33, 34, 35, 36, 37, 38, 39	
Rational functions	40, 41, 42, 43, 44	
Variation	45, 46, 47, 48	

Polynomial Equations

5

Brief Yourself

This chapter includes questions about synthetic division, finding zeros of polynomials, the factor and remainder theorems, the Fundamental Theorem of Algebra, Descartes' Rule of Signs, and complex zeros of polynomials.

Synthetic division can be used when the divisor is of the form $x \pm a$. Remember to insert zeros in the dividend for any missing powers.

Two other uses of synthetic division are:

The Remainder Theorem. $f(a)$ equals the remainder when the polynomial $f(x)$ is synthetically divided by a.

The Factor Theorem. If the remainder equals 0 when $f(x)$ is synthetically divided by a, then $x - a$ is a factor of the polynomial.

Zeros of polynomials were first introduced in Chapter 4. The real zeros are x-intercepts of the graph of the polynomial. When a polynomial is factorable, the zeros can be found by setting each factor equal to 0 and solving.

Several tests can be used to learn more about the zeros of a polynomial function. Descartes' Rule of Signs can be used to determine the number of positive and negative *real* zeros. The number of positive real zeros equals the number of sign changes (or is less than that number by an even integer). The number of negative real zeros equals the number of sign changes when x is replaced by $(-x)$ (or is less than that number by an even integer). The rational zero test provides a way of listing all the possible rational zeros (use $\pm$ factors of the constant term divided by factors of the leading coefficient). Any number in the list that gives a remainder of 0 when the polynomial is synthetically divided by that number *is* a zero. This list can be extensive, so use the Upper and Lower Bounds Rule to shorten the list. If synthetic division by $c > 0$ yields a last row of all positive numbers or zero, no larger number than c will be a zero. If synthetic division by $c < 0$ yields a last row of alternating signs, no smaller number than c is a zero.

Counting multiplicities, every polynomial of degree n has exactly n roots and exactly n linear factors. Complex zeros occur in conjugate pairs so that if $-2i$ is a zero of a given polynomial, then so is $2i$. To find all the zeros of a polynomial function, you may need to combine all the techniques from the chapter. If asked for zeros or x-intercepts, list x-values. If asked for factors, use $(x - a)$, where a is the zero, as a factor for each zero found.

Test Yourself

Divide by synthetic division.

1. $(7x^3 - x^2 + 4x + 3) \div (x - 3)$

2. $(x^4 + x^3 + x^2 + x + 1) \div (x + 4)$

3. $\dfrac{3x^5 - 2x^3 + 5}{x - 1}$

4. $\dfrac{x^6 + 1}{x + 3}$

5. Use synthetic division to find $f(2)$ and $f(-3)$ if $f(x) = x^3 + 3x^2 - 6x - 1$.

6. Use synthetic division to determine whether $x = -1$ is a zero of the function $f(x) = x^4 - 7x^3 + 11x^2 + 7x - 12$.

7. Use synthetic division to determine whether $x = -2$ is a zero of the function $f(x) = x^4 - 2x^3 - 3x^2 + 4x + 4$.

8. Use Descartes' Rule of Signs to determine the number of positive and negative zeros of $f(x) = -4x^3 - x^2 + 8x - 3$.

9. Use Descartes' Rule of Signs to determine the number of positive and negative zeros of $g(x) = -2x^6 + 8x^5 - 19x^4 + 7x^3 - 5x^2 + 2x - 1$

10. List all possible rational zeros of $f(x) = x^5 - 10x^4 + 4x^2 - 12$.

11. List all possible rational zeros of $g(x) = 3x^6 + 5x^4 + 4x^2 + 6$.

12. Find all the rational zeros of $f(x) = x^3 - x^2 - 2x + 2$.

13. Find all the rational zeros of $h(x) = -x^3 + 3x^2 - 4$.

14. Use synthetic division to verify the upper and lower bounds of the zeros of $f(x) = 3x^5 - 5x^3 + 4x^2 - 6$.

    (a) Upper: $x = 2$

    (b) Lower: $x = -2$

15. Use synthetic division to determine whether $x = 5 + 2i$ is a zero of $f(x) = x^3 - 7x^2 - x + 87$.

16. Use synthetic division to find $f(3i)$ if $f(x) = x^3 + x^2 + 9x + 9$.

17. Find a polynomial with integer coefficients that has zeros $2, 2, \dfrac{1}{3}, -\dfrac{1}{2}$.

18. Find a polynomial with integer coefficients that has zeros $3, \sqrt{2}i, -\sqrt{2}i, \dfrac{3}{4}$.

19. Use synthetic divsion:
    $$\dfrac{4x^3 + 23x^2 + 34x - 10}{x - (-3 + i)}$$

Find all the zeros of the function.

20. $f(x) = x^4 - x^3 - 11x^2 - x - 12$

21. $g(x) = x^4 - 5x^2 + 4$

22. $h(x) = 4x^3 - 8x^2 - 3x + 9$

23. $f(x) = 7x^3 + 11x^2 + 18x + 8$

24. Factor: $6x^4 - x^3 + x^2 - 5x + 2$

25. Factor: $x^4 + x^2 - 20$

Check Yourself

1. $(7x^3 - x^2 + 4x + 3) \div (x - 3)$

3	7	−1	4	3
		21	60	192
	7	20	64	195

    Change the sign of the divisor.

    $7x^2 + 20x + 64 R(195)$ The quotient is one degree less than the dividend.

    (Synthetic division)

2. $(x^4 + x^3 + x^2 + x + 1) \div (x + 4)$

−4	1	1	1	1	1
		−4	12	−52	204
	1	−3	13	−51	205

    $x^3 - 3x^2 + 13x - 51 R(205)$

    (Synthetic division)

3. $\dfrac{3x^5 - 2x^3 + 5}{x - 1}$

    Insert zeros for missing powers:

1	3	0	−2	0	0	5
		3	3	1	1	1
	3	3	1	1	1	6

    $3x^4 + 3x^3 + x^2 + x + 1 R(6)$

    (Synthetic division)

4. $\dfrac{x^6 + 1}{x + 3}$

-3	1	0	0	0	0	0	1
		-3	9	-27	81	-243	729
	1	-3	9	-27	81	-243	730

$x^5 - 3x^4 + 9x^3 - 27x^2 + 81x - 243\, R(730)$

(Synthetic division)

5. The remainder from synthetically dividing $f(x)$ by 2 gives $f(2)$:

2	1	3	-6	-1
		2	10	8
	1	5	4	7

$f(2) = 7$

The remainder from synthetically dividing $f(x)$ by -3 gives $f(-3)$:

-3	1	3	-6	-1
		-3	0	18
	1	0	-6	17

$f(-3) = 17$

(Factor and Remainder Theorems)

6. $x = -1$ is a zero if synthetic division of $f(x)$ by -1 yields a remainder of 0:

-1	1	-7	11	7	-12
		-1	8	-19	12
	1	-8	19	-12	0

Since the remainder equals 0, $x = -1$ is a zero of the function.

(Factor and Remainder Theorems)

7.

$$
\begin{array}{r|rrrrr}
-2 & 1 & -2 & -3 & 4 & 4 \\
 & & -2 & 8 & -10 & 12 \\
\hline
 & 1 & -4 & 5 & -6 & 16
\end{array}
$$

Since the remainder equals 16, and does not equal 0, $x = -2$ is not a zero of the function.

(Factor and Remainder Theorems)

8. $f(x) = -4x^3 - x^2 + 8x - 3$ has 2 sign changes. Therefore, the function has 2 or 0 positive real zeros.

$f(-x) = -4(-x)^3 - (-x)^2 + 8(-x) - 3 = 4x^3 - x^2 - 8x - 3$

$f(-x)$ has 1 sign change. Therefore, the function has 1 negative real zero.

(Zeros of polynomials)

9. $g(x) = -2x^6 + 8x^5 - 19x^4 + 7x^3 - 5x^2 + 2x - 1$ has 6 sign changes. Therefore, the function has 6 or 4 or 2 or 0 positive real zeros.

$g(-x) = -2(-x)^6 + 8(-x)^5 - 19(-x)^4 + 7(-x)^3 - 5(-x)^2 + 2(-x) - 1$

$\qquad = -2x^6 - 8x^5 - 19x^4 - 7x^3 - 5x^2 - 2x - 1$

Since $g(-x)$ has 0 sign changes, the function has 0 negative real zeros.

(Zeros of polynomials)

10. $f(x) = x^5 - 10x^4 + 4x^2 - 12$

$$\text{possible rational zeros} = \pm\left\{\frac{\text{factors of } 12}{\text{factors of } 1}\right\} = \pm\left\{\frac{1, 2, 3, 4, 6, 12}{1}\right\} = \pm\{1, 2, 3, 4, 6, 12\}$$

(Zeros of polynomials)

11. $g(x) = 3x^6 + 5x^4 + 4x^2 + 6$

$$\text{possible rational zeros} = \pm\left\{\frac{\text{factors of } 6}{\text{factors of } 3}\right\} = \pm\left\{\frac{1, 2, 3, 6}{1, 3}\right\} = \pm\left\{\frac{1}{1}, \frac{1}{3}, \frac{2}{1}, \frac{2}{3}, \frac{3}{1}, \frac{3}{3}, \frac{6}{1}, \frac{6}{3}\right\} = \pm\left\{1, \frac{1}{3}, 2, \frac{2}{3}, 3, 6\right\}$$

Note that since $\dfrac{3}{3} = 1$, and 1 is already in the list, we do not list it again. Similarly, since $2 = \dfrac{6}{3}$, and 2 is already in the list, we do not list it twice.

(Zeros of polynomials)

12. Note the difference in the directions for problems 10 and 11 versus 12. In this problem, you must actually *find* all the rational zeros, not just list the possibilities.

$f(x) = x^3 - x^2 - 2x + 2$

$$\text{possible rational zeros} = \pm\left\{\frac{\text{factors of 2}}{\text{factors of 1}}\right\} = \pm\left\{\frac{1, 2}{1}\right\} = \pm\{1, 2\}$$

Synthetically divide by each possible zero. A remainder of 0 implies the divisor *is* a zero.

1	1	−1	−2	2
		1	0	−2
	1	0	−2	0

The remainder equals 0, which means $x = 1$ is a zero. We could continue to divide into 1 −1 −2 2, but it is faster to use the bottom line 1 0 −2. A zero of $x^2 - 2$ will also be a zero of the original function. Since $x^2 - 2 = 0$ has no rational solutions, the only rational solution is $x = 1$. If you continue with the synthetic division:

2	1	0	−2
		2	4
	1	2	2

Since the remainder is not 0, 2 is not a zero.

−1	1	0	−2
		−1	1
	1	−1	−1

Since the remainder is not 0, −1 is not a zero.

−2	1	0	−2
		−2	4
	1	−2	2

Since the remainder is not 0, −2 is not a zero. Therefore, the only rational zero is $x = 1$.

(Zeros of polynomials)

13. $h(x) = -x^3 + 3x^2 - 4$

$$\text{possible rational zeros} = \pm\left\{\frac{\text{factors of 4}}{\text{factors of 1}}\right\} = \pm\left\{\frac{1, 2, 4}{1}\right\} = \pm\{1, 2, 4\}$$

$$
\begin{array}{r|rrrr}
1 & -1 & 3 & 0 & -4 \\
  &   & -1 & 2 & 2 \\
\hline
  & -1 & 2 & 2 & -2 \\
\end{array}
$$

$$
\begin{array}{r|rrrr}
2 & -1 & 3 & 0 & -4 \\
  &   & -2 & 2 & 4 \\
\hline
  & -1 & 1 & 2 & 0 \\
\end{array}
$$

Since the remainder is 0, $x = 2$ is a zero. Now use the bottom line from the division that yielded the 0.

$$
\begin{array}{r|rrr}
4 & -1 & 1 & 2 \\
  &   & -4 & -12 \\
\hline
  & -1 & -3 & -10 \\
\end{array}
$$

$$
\begin{array}{r|rrr}
-1 & -1 & 1 & 2 \\
   &   & 1 & -2 \\
\hline
   & -1 & 2 & 0 \\
\end{array}
$$

$x = -1$ is a zero. It's easier to solve the remaining equation than to continue the synthetic division:

$$-x + 2 = 0$$

$$-x = -2$$

$$x = 2$$

2 is called a zero of multiplicity 2. The rational zeros are 4, 2, and –1.

(Zeros of polynomials)

14. $f(x) = 3x^5 - 5x^3 + 4x^2 - 6$

$$
\begin{array}{r|rrrrrr}
2 & 3 & 0 & -5 & 4 & 0 & -6 \\
  &   & 6 & 12 & 14 & 36 & 72 \\
\hline
  & 3 & 6 & 7 & 18 & 36 & 66 \\
\end{array}
$$

Since the bottom line contains all positive numbers, 2 is an upper bound.

-2	3	0	-5	4	0	-6
		-6	12	-14	20	-40
	3	-6	7	-10	20	-46

Since the bottom line contains numbers that alternate in sign, -2 is a lower bound.

(Zeros of polynomials)

15. $f(x) = x^3 - 7x^2 - x + 87$

$5 + 2i$	1	-7	-1	87
		$5 + 2i$	$-14 + 6i$	-87
	1	$-2 + 2i$	$-15 + 6i$	0

$(5 + 2i)(-2 + 2i) = -10 + 10i - 4i + 4i^2 = -14 + 6i$

$(5 + 2i)(-15 + 6i) = -75 + 30i - 30i + 12i^2 = -87$

Since the remainder equals 0, $x = 5 + 2i$ is a zero of $f(x)$.

(Fundamental Theorem of Algebra and complex zeros)

16. $f(x) = x^3 + x^2 + 9x + 9$

$3i$	1	1	9	9
		$3i$	$-9 + 3i$	-9
	1	$1 + 3i$	$3i$	0

$3i(1 + 3i) = 3i + 9i^2 = 3i - 9$

$f(3i) = 0$

(Fundamental Theorem of Algebra and complex zeros)

17.   If the zeros are 2, 2, $\frac{1}{3}$, $-\frac{1}{2}$, the polynomial has factors $(x - 2)^2$, $\left(x - \frac{1}{3}\right)$, and $\left(x + \frac{1}{2}\right)$. To eliminate the

fractions, note that $\left(x - \frac{1}{3}\right)$ and $(3x - 1)$ yield the same solution of $x = \frac{1}{3}$. Therefore we can write

$f(x) = (x - 2)^2(3x - 1)(2x + 1)$

$\quad\ = (x^2 - 4x + 4)(6x^2 + x - 1)$

$$= 6x^4 - 23x^3 + 19x^2 + 8x - 4$$

(Fundamental Theorem of Algebra and complex zeros)

18. If the zeros are 3, $\sqrt{2}i$, $-\sqrt{2}i$, and $\frac{3}{4}$, the polynomial has factors $(x-3)$, $(x-\sqrt{2}i)$, $(x+\sqrt{2}i)$, and

$\left(x-\frac{3}{4}\right)$ or $(4x-3)$.

$$f(x) = (x-3)(x-\sqrt{2}i)(x+\sqrt{2}i)(4x-3).$$

Note that $(x-\sqrt{2}i)(x+\sqrt{2}i) = (x)^2 - (\sqrt{2}i)^2 = x^2 - (2i^2) = x^2 - (2(-1)) = x^2 + 2$.

Then

$$f(x) = (x-3)(4x-3)(x^2+2)$$

$$= (4x^2 - 15x + 9)(x^2 + 2)$$

$$= 4x^4 + 8x^2 - 15x^3 - 30x + 9x^2 + 18$$

$$= 4x^4 - 15x^3 + 17x^2 - 30x + 18$$

(Fundamental Theorem of Algebra and complex zeros)

19.

$-3+i$	4	23	34	-10
		$-12+4i$	$-37-i$	10
	4	$11+4i$	$-3-i$	0

$$(-3+i)(11+4i) = -33 - 12i + 11i + 4i^2 = -37 - i$$

$$(-3+i)(-3-i) = 9 + 3i - 3i - i^2 = 10$$

$$4x^2 + (11+4i)x + (-3-i)$$

(Fundamental Theorem of Algebra and complex zeros)

20. $f(x) = x^4 - x^3 - 11x^2 - x - 12$

Use Descartes' Rule of Signs to determine the number of positive and negative real zeros:

Number of sign changes in $f(x)$ is 1; there is 1 positive real zero.

$$f(-x) = (-x)^4 - (-x)^3 - 11x^2 - (-x) - 12$$

$$= x^4 + x^3 - 11x^2 + x - 12$$

The number of sign changes in $f(-x)$ is 3; there are 3 or 1 negative real zeros. List the possible rational zeros.

$$\pm\left\{\frac{\text{factors of 12}}{\text{factors of 1}}\right\} = \pm\{1, 2, 3, 4, 6, 12\}$$

Use synthetic division.

1	1	−1	−11	−1	−12
		1	0	−11	−12
	1	0	−11	−12	−24

2	1	−1	−11	−1	−12
		2	2	−18	−38
	1	1	−9	−19	−50

3	1	−1	−11	−1	−12
		3	6	−15	−48
	1	2	−5	−16	−60

4	1	−1	−11	−1	−12
		4	12	4	12
	1	3	1	3	0

Therefore, 4 is a zero. There is no need to check 6 or 12 since there is only 1 positive real zero (also, note that 4 is an upper bound since the bottom row contains all positive numbers).

Continue synthetic division using the bottom row.

−1	1	3	1	3
		−1	−2	1
	1	2	−1	4

−2	1	3	1	3
		−2	−2	2
	1	1	−1	5

$$
\begin{array}{r|rrrr}
-3 & 1 & 3 & 1 & 3 \\
 & & -3 & 0 & -3 \\
\hline
 & 1 & 0 & 1 & 0 \\
\end{array}
$$

$x = -3$ is a zero.

Although you could continue to use synthetic division, it is faster to solve $x^2 + 1 = 0$ (see bottom row of previous synthetic division).

$$x^2 + 1 = 0$$

$$x^2 = -1$$

$$x = \pm\sqrt{-1}$$

$$x = \pm i$$

The zeros are 4, –3, ± i.

(Fundamental Theorem of Algebra and complex zeros)

21. $g(x) = x^4 - 5x^2 + 4$

Note that g can be factored.

$$x^4 - 5x^2 + 4 = (x^2 - 4)(x^2 - 1) = (x - 2)(x + 2)(x - 1)(x + 1)$$

The zeros are ±2, ±1.

(Fundamental Theorem of Algebra and complex zeros)

22. $h(x) = 4x^3 - 8x^2 - 3x + 9$

Use Descartes' Rule of Signs.

The number of sign changes in $h(x)$ is 2, so there are 2 or 0 positive real zeros.

$$h(-x) = 4(-x)^3 - 8(-x)^2 - 3(-x) + 9$$

$$= -4x^3 - 8x^2 + 3x + 9$$

The number of sign changes in $h(-x)$ is 1, so there is 1 negative real zero.

Possible rational zeros:

$$\pm\left\{\frac{\text{factors of } 9}{\text{factors of } 4}\right\} = \pm\left\{\frac{1, 3, 9}{1, 2, 4}\right\} = \pm\left\{1, \frac{1}{2}, \frac{1}{4}, 3, \frac{3}{2}, \frac{3}{4}, 9, \frac{9}{2}, \frac{9}{4}\right\}$$

Since there may not be any positive real zeros, let's begin dividing by possible negative rational zeros:

$$
\begin{array}{r|rrrr}
-1 & 4 & -8 & -3 & 9 \\
   &   & -4 & 12 & -9 \\
\hline
   & 4 & -12 & 9 & 0
\end{array}
$$

The bottom row represents the quadratic $4x^2 - 12x + 9$, which can be easily factored:

$$4x^2 - 12x + 9 = (2x - 3)^2$$

So the zero $\dfrac{3}{2}$ is a zero of multiplicity 2. The zeros are -1 and $\dfrac{3}{2}$.

(Fundamental Theorem of Algebra and complex zeros)

23. $f(x) = 7x^3 + 11x^2 + 18x + 8$

There are no sign changes in $f(x)$ so there are no positive real zeros.

$f(-x) = -7x^3 + 11x^2 - 18x + 8$ has 3 sign changes so there are 3 or 1 negative real zeros.

$$\text{possible rational zeros} = -\left\{\frac{\text{factors of 8}}{\text{factors of 7}}\right\} = -\left\{\frac{1, 2, 4, 8}{1, 7}\right\} = -\left\{1, \frac{1}{7}, 2, \frac{2}{7}, 4, \frac{4}{7}, 8, \frac{8}{7}\right\}$$

$$
\begin{array}{r|rrrr}
-1 & 7 & 11 & 18 & 8 \\
   &   & -7 & -4 & -14 \\
\hline
   & 7 & 4 & 14 & -6
\end{array}
$$

$$
\begin{array}{r|rrrr}
-2 & 7 & 11 & 18 & 8 \\
   &   & -14 & 6 & -48 \\
\hline
   & 7 & -3 & 24 & -40
\end{array}
$$

Note that the signs in the bottom row alternate, which means -2 is a lower bound. No number smaller than -2 is a zero of the function.

$$
\begin{array}{r|rrrr}
-\dfrac{4}{7} & 7 & 11 & 18 & 8 \\
   &   & -4 & -4 & -8 \\
\hline
   & 7 & 7 & 14 & 0
\end{array}
$$

Now solve $7x^2 + 7x + 14 = 0$

$x^2 + x + 2 = 0$ Divide each term by 7.

$$x = \frac{-1 \pm \sqrt{1 - 4(1)(2)}}{2(1)} \qquad \text{Use the quadratic formula.}$$

$$x = \frac{-1 \pm i\sqrt{7}}{2}$$

Therefore, the zeros are $-\dfrac{4}{7}, \dfrac{-1 \pm i\sqrt{7}}{2}$

(Fundamental Theorem of Algebra and complex zeros)

24. $f(x) = 6x^4 - x^3 + x^2 - 5x + 2$. The number of sign changes is 4, so there are 4 or 2 or 0 positive real zeros. $f(-x) = 6x^4 + x^3 + x^2 + 5x + 2$. The number of sign changes in $f(-x)$ is 0, so there are 0 negative real zeros.

$$\text{possible rational zeros} = \left\{\frac{\text{factors of 2}}{\text{factors of 6}}\right\} = \left\{\frac{1, 2}{1, 2, 3, 6}\right\} = \left\{1, \frac{1}{2}, \frac{1}{3}, \frac{1}{6}, 2, \frac{2}{3}\right\}$$

1	6	−1	1	−5	2
		6	5	6	1
	6	5	6	1	3

Note that 1 is an upper bound (the bottom row contains all positive numbers), so there is no need to try 2.

$\frac{1}{2}$	6	−1	1	−5	2
		3	1	1	−2
	6	2	2	−4	0

$\frac{2}{3}$	6	2	2	−4
		4	4	4
	6	6	6	0

Use $6x^2 + 6x + 6$ to complete the factorization:

$$6x^2 + 6x + 6 = 6(x^2 + x + 1)$$

Since $x^2 + x + 1$ does not factor over the reals, find the roots using the quadratic formula:

$$x = \frac{-1 \pm \sqrt{1 - 4(1)(1)}}{2(1)} = \frac{-1 \pm \sqrt{-3}}{2} = \frac{-1 \pm i\sqrt{3}}{2}$$

The factorization is:

$$\left(x - \frac{1}{2}\right)\left(x - \frac{2}{3}\right)(6)\left(x - \left(\frac{-1 + i\sqrt{3}}{2}\right)\right)\left(x - \left(\frac{-1 - i\sqrt{3}}{2}\right)\right)$$

$$= (2x - 1)(3x - 2)\left(x - \left(\frac{-1 + i\sqrt{3}}{2}\right)\right)\left(x - \left(\frac{-1 - i\sqrt{3}}{2}\right)\right)$$

(Fundamental Theorem of Algebra and complex zeros)

25.   Note that $x^4 + x^2 - 20$ can be factored by trial and error:

$$(x^2 - 4)(x^2 + 5) = (x - 2)(x + 2)(x^2 + 5)$$

$x^2 + 5$ does not factor over the reals but can be factored over the complex numbers. Solve $x^2 + 5 = 0$:

$$x^2 = -5$$

$$x = \pm i\sqrt{5}$$

So the factors are $(x - i\sqrt{5})$ and $(x + i\sqrt{5})$.

The complete factorization is:

$$(x - 2)(x + 2)(x - i\sqrt{5})(x + i\sqrt{5})$$

(Fundamental Theorem of Algebra and complex zeros)

Grade Yourself

Circle the question numbers that you had incorrect. Then indicate the number of questions you missed. If you answered more than three questions incorrectly, you will have to focus on that topic. If a topic has fewer than three questions and you had at least one wrong, we suggest you study that topic. Read your textbook or a review book or ask your teacher for help.

Subject: Polynomial Equations

Topic	Question Numbers	Number Incorrect
Synthetic division	1, 2, 3, 4	
Factor and Remainder Theorems	5, 6, 7	
Zeros of polynomials	8, 9, 10, 11, 12, 13, 14	
Fundamental Theorem of Algebra and complex zeros	15, 16, 17, 18, 19, 20, 21, 22, 23, 24, 25	

Exponential and Logarithmic Functions

6

Brief Yourself

This chapter includes questions about graphs of exponential and logarithmic functions, properties of logarithms, and the solutions of exponential and logarithmic equations. You should be familiar with the exponent key, log and natural log keys, and e^x key on your calculator.

You should recognize the basic graphs of $y = a^x$ and $y = a^{-x}$ for $a > 1$ and should be able to shift and reflect the graphs by the same techniques used in Chapter 3. For example, the graph of $f(x) = 3^x + 1$ is the same as the graph of $g(x) = 3^x$ shifted up 1 unit.

Two formulas you may need to use involve compounding interest where P is the amount initally invested and A is the amount in the account at the end of the time period:

$A = P\left(1 + \dfrac{r}{n}\right)^{nt}$ for n compoundings per year at a rate of $r\%$ for t years

$A = Pe^{rt}$ for continuous compounding at a rate of $r\%$ for t years

Memorize the definition of logarithm to convert from logarithmic form to exponential form and vice versa:

$\log_a x = y$ if and only if $a^y = x$

Several properties of logarithms are also worth memorizing:

$\log_a 1 = 0$

$\log_a a = 1$

$\log_a a^x = x$

If $\log_a x = \log_a y$, then $x = y$

Once you recognize the basic graph of a logarithmic function, use the techniques for shifting and reflecting graphs to help you graph logarithmic functions.

Calculators usually have a common logarithm key for base 10 and a natural logarithm key for base e. To find logarithms with a different base, use the change of base formula:

Change of Base Formula: $\log_a x = \dfrac{\log x}{\log a} = \dfrac{\ln x}{\ln a}$.

There are three properties of logarithms which can be used to convert multiplication into addition, division into subtraction, and power problems into multiplication problems.

1. The log of a product is the sum of the logs: $\log_a(xy) = \log_a x + \log_a y$

2. The log of a quotient is the difference of the logs: $\log_a\left(\dfrac{x}{y}\right) = \log_a x - \log_a y$

3. The log of a number raised to a power is the product of that power and the log: $\log_a x^b = b(\log_a x)$

To solve exponential equations (equations where the variable appears as the exponent), use the following property:

If $a^x = a^y$, then $x = y$ for $a > 0$, $a \neq 1$.

Note that the bases must be equal to make use of this property. When it is not obvious how to make the bases equal, we take the logarithm of each side and solve the resulting equation. If an *exact* solution is required, leave the logarithm(s) in your answer. Otherwise, use your calculator to evaluate the logarithms and complete the solution.

To solve a logarithmic equation, we use the definition of a logarithm to write the equation in exponential form. The following steps may be used.

1. Use the properties of logarithms to write the equation with a single logarithm or in the form $\log_b N = \log_b M$.
2. If the equation can be written in the form $\log_b N = \log_b M$, then set $N = M$ and solve. Otherwise, write $\log_b x = y$ as $b^y = x$.
3. Solve the exponential equation.
4. Use only solutions for which the logarithm(s) are defined.

Test Yourself

Use your calculator to evaluate each expression. Round your results to three decimal places.

1. $(2.4)^{4.5}$

2. $3^{4\pi}$

3. $18^{\sqrt{3}}$

4. $e^{-2/3}$

Graph each exponential function.

5. $f(x) = 3^{x+1}$

6. $g(x) = e^{x/2}$

7. $h(x) = 3^{x-2} - 1$

8. A deposit of $7,500 is made to a trust fund that pays 6.5 percent interest. Find the amount of money in the account after 25 years if the money is compounded quarterly.

9. A certain type of bacteria increases according to the model

$$P(t) = 400e^{0.252t}$$

where t is the time in hours. Find the amount of bacteria after 10 hours. Round the result to one decimal place.

10. The demand equation for a certain product is given by

$$p = 4000\left(1 - \frac{3}{3 + e^{-0.001x}}\right),$$ where p is the price in dollars. Find the price p for a demand of $x = 500$ units.

11. Change to exponential form: $\log_8 4 = \frac{2}{3}$.

12. Change to logarithmic form: $4^{-3} = \frac{1}{64}$.

Evaluate the expression without a calculator.

13. $\log_6 36$

14. $\log_8 1$

15. $\log_{\sqrt{3}} \sqrt{3}$

16. $\log_5 5^4$

17. $\ln e^3$

Use a calculator to evaluate each logarithm. Round to three decimal places.

18. $\log 0.034$

19. $\ln(1 + \sqrt{5})$

20. $\ln 12^3$

Graph each logarithmic function.

21. $f(x) = \log_4 x + 2$

22. $f(x) = \log_4(x + 2)$

23. $f(x) = -\log_4 x$

Rewrite each expression as a sum or difference using the properties of logarithms.

24. $\log_2 5x$

25. $\log_{10}\left(\frac{3m}{n}\right)$

26. $\ln \sqrt{x}$

27. $\log_3 \frac{4xy}{pq}$

28. $\ln x^2 y^3$

29. $\log_2 \frac{x \sqrt[3]{y}}{z^5}$

Write each expression as a single logarithm.

30. $\log_4 x + \log_4 y$

31. $2\ln x - 3\ln z$

32. $4\log_2(x + 2)$

33. $2\log_5 x + 6\log_5 y - \frac{1}{2}\log_5 z$

34. $\frac{1}{3}\ln x - \ln y - \frac{1}{2}\ln z$

35. $3[\ln x + \ln(x + 2)] - 2\ln(x - 3)$

Solve each equation. Give all approximations to four significant digits.

36. $8^x = 4$

37. $3^x = \frac{1}{81}$

38. $9^{x+2} = 27^{x-1}$

39. $5^x = 12$

40. $4e^{2x} = \sqrt{5}$

41. $\log_6 x + \log_6(x - 1) = 1$

42. $\log 2 + \log(5x - 2) = \log(8x + 2)$

43. $\ln 2x = 6.8$

44. $\ln(3x - 1) = \ln x + \ln 4$

45. $\ln x + \ln(x - 4) = 1$

Check Yourself

1.   51.399

   (Exponential functions)

2.   990,107.875

   (Exponential functions)

3.   149.347

   (Exponential functions)

4.   0.513

   (Exponential functions)

5. $f(x) = 3^{x+1}$

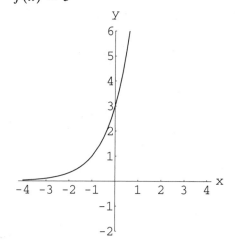

   (Exponential functions)

6. $g(x) = e^{x/2}$

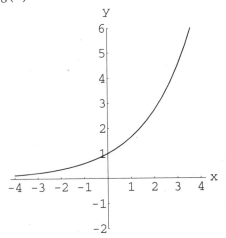

(Exponential functions)

7. $h(x) = 3^{x-2} - 1$

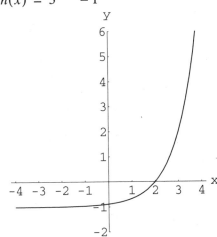

(Exponential functions)

8. Use the formula $A = P\left(1 + \dfrac{r}{n}\right)^{nt}$ where $P = \$7{,}500$, $r = 0.065$, $n = 4$ (because it is compounded 4 times per year) and $t = 25$ years.

$$A = 7500\left(1 + \frac{0.065}{4}\right)^{4(25)}$$

$$A = 7500(1.01625)^{100}$$

$$A = \$37{,}593.88$$

(Exponential functions)

9. $P(t) = 400e^{0.252t}$ and $t = $ hours

$P(t) = 400e^{0.252(10)} = 400e^{2.52} \approx 4971.4$ bacteria.

(Exponential functions)

10. $p = 4000\left(1 - \dfrac{3}{3 + e^{-0.001x}}\right)$ and $x = 500$.

$p = 4000\left(1 - \dfrac{3}{3 + e^{-0.001(500)}}\right)$ Substitute $x = 500$.

$\quad = 4000\left(1 - \dfrac{3}{3.6065\ldots}\right)$

$\quad \approx \$672.70$

(Exponential functions)

11. $\log_8 4 = \dfrac{2}{3} \Leftrightarrow 8^{2/3} = 4$.

(Logarithmic functions)

12. $4^{-3} = \dfrac{1}{64} \Leftrightarrow \log_4 \dfrac{1}{64} = -3$.

(Logarithmic functions)

13. $\log_6 36 = x$ Set the expression equal to x.

$\quad 6^x = 36$ Convert to exponential form.

$\quad 6^x = 6^2$ Make the bases equal.

$\quad x = 2$ Use $x^a = x^x \Rightarrow a = b$.

(Logarithmic functions)

14. $\log_8 1 = 0$ Use $\log_a 1 = 0$.

(Logarithmic functions)

15. $\log_{\sqrt{3}} \sqrt{3} = 1$ Use $\log_a a = 1$.

(Logarithmic functions)

16. $\log_5 5^4 = 4$ Use $\log_a a^x = x$.

(Logarithmic functions)

17. $\ln e^3 = \ln_e e^3 = 3$ Use $\log_a a^x = x$.

(Logarithmic functions)

18. $\log 0.034 \approx -1.469$

(Logarithmic functions)

19. $\ln(1 + \sqrt{5}) \approx 1.174$

Calculate $1 + \sqrt{5}$ first, then find the natural log.

(Logarithmic functions)

20. $\ln 12^3 \approx 7.455$. Calculate 12^3 first, then find the natural log. Or write $\ln 12^3 = 3 \ln 12$ using Property 3 for logarithms, calculate $\ln 12$, then multiply by 3.

(Logarithmic functions)

21. $f(x) = \log_4 x + 2$

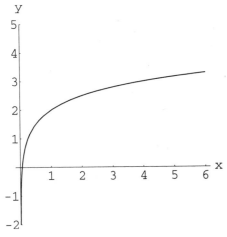

(Logarithmic functions)

22. $f(x) = \log_4 (x + 2)$

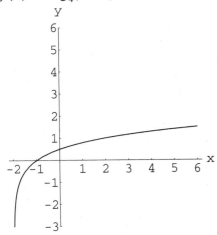

(Logarithmic functions)

23. $f(x) = -\log_4 x$

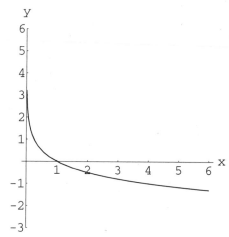

Notice that each graph has the same shape as $y = \log_4 x$, but $f(x) = \log_4 x + 2$ is shifted up 2 units, $f(x) = \log_4(x+2)$ is shifted left 2 units, and $f(x) = -\log_4 x$ is reflected through the x-axis.

(Logarithmic functions)

24. $\log_2 5x = \log_2 5 + \log_2 x$

(Properties of logarithms)

25. $\log_{10} \dfrac{3m}{n} = \log_{10} 3m - \log_{10} n$

$\qquad = \log_{10} 3 + \log_{10} m - \log_{10} n$

(Properties of logarithms)

26. $\ln \sqrt{x} = \ln x^{1/2} = \dfrac{1}{2}\ln x$

(Properties of logarithms)

27. $\log_3 \dfrac{4xy}{pq} = \log_3(4xy) - \log_3(pq)$

$\qquad = \log_3 4 + \log_3 x + \log_3 y - (\log_3 p + \log_3 q)$

$\qquad = \log_3 4 + \log_3 x + \log_3 y - \log_3 p - \log_3 q$

(Properties of logarithms)

28. $\ln x^2 y^3 = \ln x^2 + \ln y^3$

$\qquad = 2\ln x + 3\ln y$

(Properties of logarithms)

29. $\log_2 \dfrac{x\sqrt[3]{y}}{z^5} = \log_2(xy^{1/3}) - \log_2 z^5$

$$= \log_2 x + \log_2 y^{1/3} - \log_2 z^5$$

$$= \log_2 x + \frac{1}{3}\log_2 y - 5\log_2 z$$

(Properties of logarithms)

30. $\log_4 x + \log_4 y = \log_4 xy$

(Properties of logarithms)

31. $2\ln x - 3\ln z = \ln x^2 - \ln z^3 = \ln \dfrac{x^2}{z^3}$

(Properties of logarithms)

32. $4\log_2(x+2) = \log_2(x+2)^4$

Note that $\log_2(x+2) \neq (\log_2 x)(\log_2 2)$.

(Properties of logarithms)

33. $2\log_5 x + 6\log_5 y - \dfrac{1}{2}\log_5 z = \log_5 x^2 + \log_5 y^6 - \log_5 z^{1/2}$

$$= \log_5 x^2 y^6 - \log_5 z^{1/2}$$

$$= \log_5 \frac{x^2 y^6}{z^{1/2}} \text{ or } \log_5 \frac{x^2 y^6}{\sqrt{z}}$$

(Properties of logarithms)

34. $\dfrac{1}{3}\ln x - \ln y - \dfrac{1}{2}\ln z$

$= \ln x^{1/3} - \ln y - \ln z^{1/2}$ Use $b\ln x = \ln x^b$.

$= \ln x^{1/3} - (\ln y + \ln z^{1/2})$ Factor out -1.

$= \ln x^{1/3} - (\ln yz^{1/2})$ Use $\ln x + \ln y = \ln xy$.

$= \ln \dfrac{x^{1/3}}{yz^{1/2}}$ Use $\ln x - \ln y = \ln \dfrac{x}{y}$.

$$= \ln \frac{\sqrt[3]{x}}{y\sqrt{z}}$$

Convert fractional exponents to radical form.

(Properties of logarithms)

35. $3[\ln x + \ln(x+2)] - 2\ln(x-3)$

$= 3[\ln x(x+2)] - 2\ln(x-3)$

Use $\ln a + \ln b = \ln ab$.

$= \ln[x(x+2)]^3 - \ln(x-3)^2$

Use $b \ln x = \ln x^b$.

$= \ln \frac{[x(x+2)]^3}{(x-3)^2}$

Use $\ln a - \ln b = \ln \frac{a}{b}$.

$= \ln \frac{x^3(x+2)^3}{(x-3)^2}$

Use $(xy)^a = x^a y^a$.

(Properties of logarithms)

36. $8^x = 4$

$(2^3)^x = 2^2$

Make the bases equal.

$2^{3x} = 2^2$

Use $(x^a)^b = x^{ab}$.

$3x = 2$

Since the bases are equal, set the exponents equal.

$x = \frac{2}{3}$

Solve for x.

(Exponential and logarithmic equations)

37. $3^x = \frac{1}{81}$

$3^x = \frac{1}{3^4}$

$81 = 3^4$.

$3^x = 3^{-4}$

Make the bases equal.

$x = -4$

Since the bases are equal, set the exponents equal.

(Exponential and logarithmic equations)

38. $9^{x+2} = 27^{x-1}$

$(3^2)^{x+2} = (3^3)^{x-1}$

Factor to make the bases equal.

$3^{2x+4} = 3^{3x-3}$

Use $(x^a)^b = x^{ab}$.

$2x+4 = 3x-3$

Since the bases are equal, set the exponents equal.

$$7 = x \qquad \text{Solve for } x.$$

(Exponential and logarithmic equations)

39. $\quad 5^x = 12$

$\log 5^x = \log 12 \qquad$ Since the bases cannot be made equal, take the log of both sides.

$x\log 5 = \log 12 \qquad$ Use $\log x^a = a\log x$.

$x = \dfrac{\log 12}{\log 5} \qquad$ Divide both sides by log5.

If an exact answer is needed, use $x = \dfrac{\log 12}{\log 5}$. If you are using a calculator for an approximation, $x \approx 1.544$ (to four significant digits).

(Exponential and logarithmic equations)

40. $\quad 4e^{2x} = \sqrt{5}$

$e^{2x} = \dfrac{\sqrt{5}}{4}$

$\ln e^{2x} = \ln\dfrac{\sqrt{5}}{4} \qquad$ Take the natural log of both sides.

$2x\ln e = \ln\dfrac{\sqrt{5}}{4} \qquad$ Use $\ln x^b = b\ln x$.

$2x = \ln\dfrac{\sqrt{5}}{4} \qquad$ $\ln e = 1.$

$x = \dfrac{1}{2}\ln\dfrac{\sqrt{5}}{4} \qquad$ This is the exact answer.

$x \approx -0.2908 \qquad$ Rounded to four significant digits.

(Exponential and logarithmic equations)

41. $\quad \log_6 x + \log_6(x-1) = 1$

$\log_6 x(x-1) = 1 \qquad$ Use $\log_a x + \log_a y = \log_a xy$.

$6^1 = x(x-1) \qquad$ Convert to exponential form.

$6 = x^2 - x \qquad$ Multiply.

$x^2 - x - 6 = 0 \qquad$ Solve the quadratic equation.

$$(x-3)(x+2) = 0$$

$x = 3$ or $x = -2$ Set each factor equal to 0 and solve.

If $x = -2$, $\log_6 x$ is not defined, so -2 cannot be used as a solution.

If $x = 3$, $\log_6 3 + \log_6(3-1) = \log_6 3(2) = \log_6 6 = 1$, so 3 does check.

The solution is $x = 3$.

(Exponential and logarithmic equations)

42. $\log 2 + \log(5x-2) = \log(8x+2)$

         $\log 2(5x-2) = \log(8x+2)$ Use $\log_a x + \log_a y = \log_a xy$.

             $2(5x-2) = 8x+2$ If $\log_a M = \log_a N$ then $M = N$.

               $10x - 4 = 8x+2$ Solve.

                     $2x = 6$

                       $x = 3$

(Exponential and logarithmic equations)

43. $\ln 2x = 6.8$

        $e^{6.8} = 2x$ Convert to exponential form.

        $\dfrac{1}{2}e^{6.8} = x$ Solve for x.

          $x \approx 448.9$

(Exponential and logarithmic equations)

44. $\ln(3x-1) = \ln x + \ln 4$

        $\ln(3x-1) = \ln 4x$ Use $\ln a + \ln b = \ln ab$.

           $3x - 1 = 4x$ If $\ln M = \ln N$, then $M = N$.

              $-1 = x$ Solve for x.

But, if $x = -1$, $\ln x$ is not defined. Therefore, there is no solution.

(Exponential and logarithmic equations)

45. $\ln x + \ln(x-4) = 1$

        $\ln x(x-4) = 1$ Use $\ln a + \ln b = \ln ab$.

          $e^1 = x(x-4)$ Convert to exponential form.

$$e = x^2 - 4x \qquad \text{Multiply.}$$

$$x^2 - 4x - e = 0 \qquad \text{Solve the quadratic equation.}$$

$$x = \frac{-(-4) \pm \sqrt{(-4)^2 - 4(1)(-e)}}{2(1)} \qquad \text{Use the quadratic formula.}$$

$$x = \frac{4 \pm \sqrt{16 + 4e}}{2}$$

$$x = \frac{4 \pm \sqrt{4(4 + e)}}{2} \qquad \text{Factor.}$$

$$x = \frac{4 \pm 2\sqrt{4 + e}}{2} \qquad \text{Use } \sqrt{4} = 2.$$

$$x = \frac{2(2 \pm \sqrt{4 + e})}{2} \qquad \text{Factor out 2.}$$

$$x = 2 \pm \sqrt{4 + e} \qquad \text{Reduce.}$$

$$x \approx 4.592 \text{ or } -0.5920$$

Note that x cannot be negative (or $\ln x$ will not be defined), so $x = 2 + \sqrt{4 + e}$ exactly or $x = 4.592$.

(Exponential and logarithmic equations)

Grade Yourself

Circle the question numbers that you had incorrect. Then indicate the number of questions you missed. If you answered more than three questions incorrectly, you will have to focus on that topic. If a topic has fewer than three questions and you had at least one wrong, we suggest you study that topic. Read your textbook or a review book or ask your teacher for help.

Subject: Exponential and Logarithmic Functions

Topic	Question Numbers	Number Incorrect
Exponential functions	1, 2, 3, 4, 5, 6, 7, 8, 9, 10	
Logarithmic functions	11, 12, 13, 14, 15, 16, 17, 18, 19, 20, 21, 22, 23	
Properties of logarithms	24, 25, 26, 27, 28, 29, 30, 31, 32, 33, 34, 35	
Exponential and logarithmic equations	36, 37, 38, 39, 40, 41, 42, 43, 44, 45	

Conic Sections

7

This chapter contains questions about parabolas, ellipses, hyperbolas, and the general second-degree equation in two variables. Circles were previously discussed in Chapter 3. Parabolas as functions were discussed in Chapter 4, but will be discussed here in more general terms.

The standard equations and formulas for parabolas are:

$(x - h)^2 = 4p(y - k)$

    Opening: up if $p > 0$, down if $p < 0$

    Vertex: (h, k)

    Focus: $(h, k + p)$

    Directrix: $y = k - p$

$(y - k)^2 = 4p(x - h)$

    Opening: right if $p > 0$, left if $p < 0$

    Vertex: (h, k)

    Focus: $(h + p, k)$

    Directrix: $x = h - p$

Note that $|p|$ = the distance from the vertex to the focus = the distance from the vertex to the directrix. To find the standard form of a parabola, you may need to use the process of completing the square. Be prepared to find the vertex, focus and/or directrix given an equation of a parabola and to find the equation of a parabola given various information about the parabola.

The standard equations and formulas for ellipses are, for $a > b$:

Major axis horizontal:

$$\frac{(x-h)^2}{a^2} + \frac{(y-k)^2}{b^2} = 1$$

Center: (h, k)

Vertices: $(h + a, k)$ and $(h - a, k)$

Foci: $(h + c, k)$ and $(h - c, k)$, where $c^2 = a^2 - b^2$

Major axis vertical:

$$\frac{(x-h)^2}{b^2} + \frac{(y-k)}{a^2} = 1$$

Center: (h, k)

Vertices: $(h, k + a)$ and $(h, k - a)$

Foci: $(h, k + c)$ and $(h, k - c)$, where $c^2 = a^2 - b^2$

The larger denominator determines whether the major axis is horizontal or vertical. Note that the foci lie on the major axis c units from the center. Be prepared to graph an ellipse, find the equation of an ellipse, or find the center, vertices, or foci given the equation of an ellipse. In the standard form of an ellipse, 1 must be isolated on the right side of the equation and the coefficients of the squared terms must equal 1.

The standard equations and formulas for hyperbolas are:

Opening left and right:

$$\frac{(x-h)^2}{a^2} - \frac{(y-k)^2}{b^2} = 1$$

Center: (h, k)

Vertices: $(h \pm a, k)$

Foci: $(h \pm c, k)$, where $c^2 = a^2 + b^2$

Asymptotes: $y - k = \pm\frac{b}{a}(x - h)$

Opening up and down:

$$\frac{(y-k)^2}{b^2} - \frac{(x-h)^2}{a^2} = 1$$

Center: (h, k)

Vertices: $(h, k \pm b)$

Foci: $(h, k \pm c)$, where $c^2 = a^2 + b^2$

Asymptotes: $y - k = \pm \frac{b}{a}(x - h)$

When the equation of a hyperbola is written in standard form and the $(x-h)^2$ term is positive, the hyperbola opens left and right. When the equation is in standard form and $(y-k)^2$ is positive, the hyperbola opens up and down.

To graph a hyperbola:

1. Draw the fundamental rectangle. This rectangle has center (h, k). Put points left and right a units from the center and up and down b units from the center. These four points are midpoints on the sides of the fundamental rectangle.
2. Draw the asymptotes of the hyperbola. These asymptotes contain the diagonals of the fundamental rectangle. Sketch the hyperbola so that it contains the appropriate vertices and approaches the asymptotes.

In the standard form of a hyperbola, 1 must be isolated on the right side of the equation and the coefficients of the squared terms must equal 1.

If only the type of graph is required, rather than completing the square to put the equation in standard form (if it is not already in standard form), use the following:

The graph of $Ax^2 + Cy^2 + Dx + Ey + F = 0$ is:

A parabola if A or C is zero.

A circle if $A = C$.

An ellipse if A and C have the same sign, but $A \neq C$.

A hyperbola if A and C have opposite signs.

Note that these conditions do not include any rotations or degenerate forms.

Test Yourself

Find the vertex, focus, and equation of the directrix for each parabola.

1. $(x-3)^2 = 8(y-1)$

2. $y = x^2 - 4x + 8$

3. $y = 2x^2 - 16x + 8$

4. $y = -x^2 - 2x - 3$

5. $(y+2)^2 = 2(x+4)$

6. $x = -2(y-1)^2 + 3$

7. $x = 3y^2 + 6y + 5$

8. Find an equation of the parabola with vertex $(0, 0)$ and focus $(0, -\frac{1}{4})$.

9. Find an equation of the parabola with vertex $(0, 0)$ and focus $(-3, 0)$.

10. Find an equation of the parabola with vertex $(3, 2)$ and directrix $y = 0$.

11. Find an equation of the parabola passing through the point $(-1, 3)$ with vertex $(-2, 1)$ and opening up.

12. A satellite dish with a parabolic cross section has a radius of 5 feet and is 3 feet deep at its center. How far is the focus from the center?

Find the center and vertices of the ellipse and sketch its graph.

13. $\frac{x^2}{16} + \frac{y^2}{9} = 1$

14. $9x^2 + 4y^2 = 36$

15. $\frac{(x-2)^2}{9} + \frac{(y-1)^2}{4} = 1$

16. $x^2 + 2y^2 - 2x - 4y = 1$

17. Find the equation of the ellipse with center at the origin, foci at $(\pm 2, 0)$, and length of minor axis 1.

18. Find the equation of the ellipse with center at the origin, foci at $(\pm 4, 0)$, and vertices at $(\pm 6, 0)$.

19. Find the equation of the ellipse with vertices $(2, 0)$ and $(2, 4)$ and minor axis of length 2.

20. Find the equation of the ellipse with foci $(0, 0)$ and $(0, 8)$ and major axis of length 16.

Find the center, vertices, and foci of the hyperbola. Sketch its graph, including the asymptotes.

21. $\frac{x^2}{9} - \frac{y^2}{16} = 1$

22. $\frac{y^2}{25} - \frac{x^2}{4} = 1$

23. $4x^2 - 49y^2 = 196$

24. $\frac{(y-1)^2}{4} - \frac{(x+2)^2}{1} = 1$

25. $3x^2 - 2y^2 - 6x - 8y - 11 = 0$

26. Find an equation for the hyperbola with center at the origin, vertices at $(\pm 2, 0)$, and foci at $(\pm 4, 0)$.

27. Find an equation for the hyperbola with center at the origin, vertices at $(0, \pm 3)$, and asymptotes $y = \pm 3x$.

28. Find an equation for the hyperbola with vertices at $(2, 0)$ and $(6, 0)$ and foci at $(-1, 0)$ and $(9, 0)$.

29. Find an equation for the hyperbola with vertices at $(3, 2)$ and $(-3, 2)$ and passing through the point $(5, 0)$.

Identify each of the following as the graph of a parabola, circle, ellipse, or hyperbola.

30. $4y^2 + 2x - 2y + 8 = 0$

31. $6x^2 + 6y^2 - 2x + 3y - 6 = 0$

32. $2x^2 - 4y^2 + 4y - 10 = 0$

33. $-3x^2 - 3y^2 + 2x - 4y + 8 = 0$

34. $-2x^2 - 4y^2 + 8x + 12 = 0$

Check Yourself

1. $(x-3)^2 = 8(y-1)$ is in the form $(x-h)^2 = 4p(y-k)$ where $h = 3, k = 1$, and $4p = 8$ so $p = 2$, and the parabola opens up.

    Vertex: $(h, k) = (3, 1)$

    Focus: $(h, k+p) = (3, 1+2) = (3, 3)$

    Directrix: $y = k-p$ so $y = 1-2$ or $y = -1$

    (Parabolas)

2. $\qquad y = x^2 - 4x + 8$

    $\qquad y - 8 = x^2 - 4x$ $\hspace{4cm}$ Subtract 8 from each side.

    $\qquad y - 8 + 4 = x^2 - 4x + 4$ $\hspace{2cm}$ $\left(\frac{1}{2} \cdot -4\right)^2$ Add 4 to each side.

    $\qquad y - 4 = (x-2)^2$ $\hspace{4cm}$ Factor.

    $\qquad (x-2)^2 = 1(y-4)$ $\hspace{3cm}$ Write in $(x-h)^2 = 4p(y-k)$ form.

    $h = 2, k = 4, 4p = 1$ so $p = \frac{1}{4}$, and the parabola opens up.

    Vertex: $(2, 4)$

    Focus: $\left(2, 4+\frac{1}{4}\right) = \left(2, \frac{17}{4}\right)$

    Directrix: $y = 4 - \frac{1}{4}$ so $y = \frac{15}{4}$

    (Parabolas)

3. $\qquad y = 2x^2 - 16x + 8$

    $\qquad y - 8 = 2(x^2 - 8x)$ $\hspace{3cm}$ Subtract 8 from each side. Factor out 2.

    $\qquad y - 8 + 32 = 2(x^2 - 8x + 16)$ $\hspace{1cm}$ $\left(\frac{1}{2} \cdot 8\right)^2 = 16$. Add 2(16) to each side.

    $\qquad y + 24 = 2(x-4)^2$ $\hspace{3cm}$ Add like terms. Factor.

    $\qquad \frac{1}{2}(y+24) = (x-4)^2$ $\hspace{3cm}$ Solve for $(x-h)^2$.

$$(x-4)^2 = \frac{1}{2}(y+24)$$ Write in standard form.

$h = 4, k = -24$, and $4p = \frac{1}{2}$ so $p = \frac{1}{8}$.

The parabola opens up.

Vertex: $(4, -24)$

Focus: $\left(4, -24+\frac{1}{8}\right) = \left(4, \frac{-191}{8}\right)$

Directrix: $y = -24 - \frac{1}{8}$ or $y = -\frac{193}{8}$

(Parabolas)

4. $y = -x^2 - 2x - 3$

$y + 3 = -(x^2 + 2x)$ Add 3 to each side. Factor out -1.

$y + 3 - 1 = -(x^2 + 2x + 1)$ $\left(\frac{1}{2} \cdot 2\right)^2 = 1$. Add $-1(1)$ to each side.

$y + 2 = -(x+1)^2$ Factor.

$(x+1)^2 = -1(y+2)$ Write in standard form.

$h = -1, k = -2, 4p = -1$ so $p = -\frac{1}{4}$.

Vertex: $(-1, -2)$

Focus: $\left(-1, -2-\frac{1}{4}\right) = \left(-1, -\frac{9}{4}\right)$

Directrix: $y = -2 - \left(-\frac{1}{4}\right)$ so $y = -\frac{7}{4}$

(Parabolas)

5. $(y+2)^2 = 2(x+4)$ is in the form $(y-k)^2 = 4p(x-h)$ where $h = -4, k = -2, 4p = 2$ so $p = \frac{1}{2}$.

The parabola opens to the right.

Vertex: $(h, k) = (-4, -2)$

Focus: $(h+p, k) = \left(-4+\frac{1}{2}, -2\right) = \left(-\frac{7}{2}, -2\right)$

Directrix: $x = h - p$

$$x = -4 - \frac{1}{2}$$

$$x = \frac{-9}{2}$$

(Parabolas)

6. $\qquad x = -2(y-1)^2 + 3$ $\qquad\qquad$ Isolate $(y-k)^2$.

$$(x-3) = -2(y-1)^2$$

$$-\frac{1}{2}(x-3) = (y-1)^2$$

$$(y-1)^2 = -\frac{1}{2}(x-3)$$

$h = 3, k = 1, 4p = -\frac{1}{2}$ so $p = -\frac{1}{8}$.

The parabola opens to the left.

Vertex: $(3, 1)$

Focus: $\left(3 - \frac{1}{8}, 1\right) = \left(\frac{23}{8}, 1\right)$

Directrix: $x = h - p$

$$x = 3 - \left(-\frac{1}{8}\right)$$

$$x = \frac{25}{8}$$

(Parabolas)

7. $\qquad x = 3y^2 + 6y + 5$

$$x - 5 = 3(y^2 + 2y)$$

$$x - 5 + 3 = 3(y^2 + 2y + 1) \qquad\qquad \left(\frac{1}{2} \cdot 2\right)^2 = 1. \text{ Add } 3(1) \text{ to each side.}$$

$$x - 2 = 3(y+1)^2 \qquad\qquad\qquad \text{Isolate } (y-k)^2.$$

$$\frac{1}{3}(x-2) = (y+1)^2$$

$$(y+1)^2 = \frac{1}{3}(x-2)$$

$h = 2, k = -1, 4p = \dfrac{1}{3}$ so $p = \dfrac{1}{12}$.

Vertex: $(2, -1)$

Focus: $\left(2 + \dfrac{1}{12}, -1\right) = \left(\dfrac{25}{12}, -1\right)$

Directrix: $x = h - p$

$$x = 2 - \dfrac{1}{12}$$

$$x = \dfrac{23}{12}$$

(Parabolas)

8. Vertex $(0, 0)$ and focus $\left(0, -\dfrac{1}{4}\right)$.

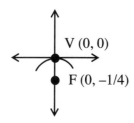

A quick sketch reveals that the parabola opens down, so the standard form is $(x - h)^2 = 4p(y - k)$ where $(h, k) = (0, 0)$ and $p = -\dfrac{1}{4}$.

$$(x - 0)^2 = 4\left(-\dfrac{1}{4}\right)(y - 0)$$

$$x^2 = -y$$

(Parabolas)

9. Vertex $(0, 0)$ and focus $(-3, 0)$.

F $(-3, 0)$ V $(0, 0)$

A quick sketch reveals that the parabola opens to the left, so the standard form is $(y - k)^2 = 4p(x - h)$ where $(h, k) = (0, 0)$ and $p = -3$.

$$(y-0)^2 = 4(-3)(x-0)$$

$$y^2 = -12x$$

(Parabolas)

10.   Vertex (3, 2) and directrix $y = 0$.

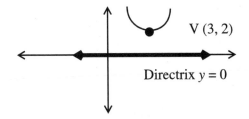

V (3, 2)

Directrix $y = 0$

A quick sketch reveals that the parabola opens up, so the standard form is $(x-h)^2 = 4p(y-k)$, where $(h, k) = (3, 2)$ and $p = 2$.

$$(x-3)^2 = 4(2)(y-2)$$

$$(x-3)^2 = 8(y-2)$$

(Parabolas)

11.   Point (−1, 3) and vertex (−2, 1).

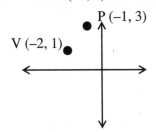

P (−1, 3)

V (−2, 1)

Since the parabola opens up, the standard form is $(x-h)^2 = 4p(y-k)$.

Use $(h, k) = (-2, 1)$, and $(x, y) = (-1, 3)$ to substitute into standard form:

$$(-1-(-2))^2 = 4p(3-1) \qquad \text{Substitute.}$$

$$1 = 4p(2) \qquad \text{Solve for } p.$$

$$\frac{1}{8} = p$$

Therefore, the equation is $(x+2)^2 = 4\left(\frac{1}{8}\right)(y-1)$ or $(x+2)^2 = \frac{1}{2}(y-1)$.

(Parabolas)

12.

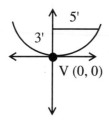

Let the vertex be at the origin, so $(h, k) = (0, 0)$. Then a point on the parabola is $(5, 3)$.

$(x - h)^2 = 4p(y - k)$ Standard form.

$(5 - 0)^2 = 4p(3 - 0)$ Substitute $(h, k) = (0, 0)$ and $(x, y) = (5, 3)$.

$25 = 12p$

$\dfrac{25}{12} = p$ Solve for p.

The focus is thus $\dfrac{25}{12}$ feet from the vertex, which is at the center of the dish.

(Parabolas)

13.   The major axis is horizontal since the larger number is in the denominator of the x^2 term.

$(h, k) = (0, 0)$, $a^2 = 16$ so $a = 4$, $b^2 = 9$ so $b = 3$.

Center: $(h, k) = (0, 0)$

Vertices: $(h + a, k) = (0 + 4, 0) = (4, 0)$

$(h - a, k) = (0 - 4, 0) = (-4, 0)$

Graph:

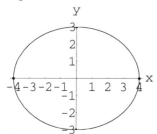

(Ellipses)

14. $\dfrac{9x^2}{36} + \dfrac{4y^2}{36} = \dfrac{36}{36}$ Divide each term by 36.

$\dfrac{x^2}{4} + \dfrac{y^2}{9} = 1$ Write in standard form.

The major axis is vertical, $(h, k) = (0, 0)$, $a^2 = 9$ so $a = 3$, $b^2 = 4$ so $b = 2$.

Center: $(h, k) = (0, 0)$

Vertices: $(h, k + a) = (0, 0 + 3) = (0, 3)$

$(h, k - a) = (0, 0 - 3) = (0, -3)$

Graph:

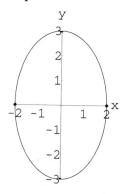

(Ellipses)

15. $\dfrac{(x - 2)^2}{9} + \dfrac{(y - 1)^2}{4} = 1$

$(h, k) = (2, 1)$, $a^2 = 9$, so $a = 3$, $b^2 = 4$, so $b = 2$.

Center: $(h, k) = (2, 1)$

Vertices: $(h + a, k) = (2 + 3, 1) = (5, 1)$

$(h - a, k) = (2 - 3, 1) = (-1, 1)$

Graph:

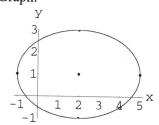

(Ellipses)

16. $\qquad x^2 + 2y^2 - 2x - 4y = 1$

$x^2 - 2x + 1 + 2(y^2 - 2y) = 1 + 1$ Complete the square on x's.

$(x - 1)^2 + 2(y^2 - 2y + 1) = 1 + 1 + 2$ Complete the square on y's.

$\qquad (x - 1)^2 + 2(y - 1)^2 = 4$

$$\frac{(x-1)^2}{4} + \frac{2(y-1)^2}{4} = \frac{4}{4} \qquad \text{Divide each term by 4.}$$

$$\frac{(x-1)^2}{4} + \frac{(y-1)^2}{2} = 1$$

Since the larger number is in the denominator of the x term, the major axis is horizontal.

$(h, k) = (1, 1)$, $a^2 = 4$, so $a = 2$, $b^2 = 2$, so $b = \sqrt{2}$.

Center: $(h, k) = (1, 1)$

Vertices: $(h + a, k) = (1 + 2, 1) = (3, 1)$

$\qquad\qquad (h - a, k) = (1 - 2, 1) = (-1, 1)$

Graph:

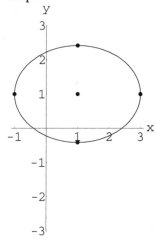

(Ellipses)

17. Center at the origin, foci at $(\pm 2, 0)$, length of minor axis 1.

A quick sketch reveals that the major axis is horizontal, $(h, k) = (0, 0)$.

$c = 2$ so $c^2 = 4$.

$2b = 1$ (length of minor axis is $2b$).

$b = \dfrac{1}{2}$

$c^2 = a^2 - b^2$

$4 = a^2 - \dfrac{1}{4}$

$4 + \dfrac{1}{4} = a^2$

$$\frac{17}{4} = a^2$$

The standard form is:

$$\frac{(x-0)^2}{17/4} + \frac{(y-0)^2}{1/4} = 1$$

$$\frac{x^2}{17/4} + \frac{y^2}{1/4} = 1$$

(Ellipses)

18.   Center at the origin, foci at $(\pm4, 0)$, vertices at $(\pm6, 0)$.

A quick sketch reveals that the major axis is horizontal.

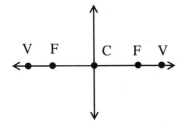

$(h, k) = (0, 0)$.

$c = 4$ so $c^2 = 16$.

$a^2 = 6^2$

$c^2 = a^2 - b^2$

$16 = 36 - b^2$

$-20 = -b^2$

$20 = b^2$

The standard form is:

$$\frac{x^2}{36} + \frac{y^2}{20} = 1$$

(Ellipses)

19.   Vertices $(2, 0)$ and $(2, 4)$ and minor axis of length 2. The center must be halfway between the vertices, so $(h, k) = (2, 2)$ and the ellipse has a vertical major axis. A minor axis of length 2 means $2b = 2$ so $b = 1$.

The distance from the center to the vertex is 2, so $a = 2$.

$$\frac{(x-h)^2}{b^2} + \frac{(y-k)^2}{a^2} = 1$$

$$\frac{(x-2)^2}{1} + \frac{(y-2)^2}{4} = 1$$

(Ellipses)

20. Foci $(0, 0)$ and $(0, 8)$ and major axis of length 16.

The center (h, k) is halfway between the foci so $(h, k) = (0, 4)$ and the major axis is vertical.

The length of the major axis is $2a = 16$, so $a = 8$.

$$c^2 = a^2 - b^2$$

$$16 = 64 - b^2$$

$$-48 = -b^2$$

$$48 = b^2$$

$$\frac{(x-0)^2}{48} + \frac{(y-4)^2}{64} = 1 \text{ or } \frac{x^2}{48} + \frac{(y-4)^2}{64} = 1$$

(Ellipses)

21. $\dfrac{x^2}{9} - \dfrac{y^2}{16} = 1$

The hyperbola opens left and right.

$(h, k) = (0, 0)$, $a^2 = 9$, $b^2 = 16$.

$$c^2 = a^2 + b^2$$

$$c^2 = 9 + 16$$

$$c^2 = 25$$

Center: $(h, k) = (0, 0)$

Vertices: $(h \pm a, k) = (0 \pm 3, 0) = (\pm 3, 0)$

Foci: $(h \pm c, k) = (0 \pm 5, 0) = (\pm 5, 0)$

Graph:

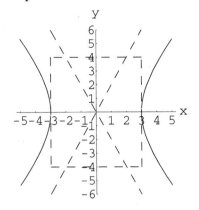

(Hyperbolas)

22. $\dfrac{y^2}{25} - \dfrac{x^2}{4} = 1$

The hyperbola opens up and down.

$(h, k) = (0, 0)$, $a^2 = 4$, $b^2 = 25$.

$c^2 = a^2 + b^2$

$c^2 = 4 + 25$

$c^2 = 29$

Center: $(h, k) = (0, 0)$

Vertices: $(h, k \pm b) = (0, 0 \pm 5) = (0, \pm 5)$

Foci: $(h, k \pm c) = (0, 0 \pm \sqrt{29}) = (0, \pm \sqrt{29})$

Graph:

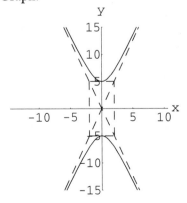

(Hyperbolas)

23. $4x^2 - 49y^2 = 196$

$$\frac{4x^2}{196} - \frac{49y^2}{196} = \frac{196}{196}$$ Divide each term by 196.

$$\frac{x^2}{49} - \frac{y^2}{4} = 1$$

The hyperbola opens left and right.

$(h, k) = (0, 0)$. $a^2 = 49, b^2 = 4$.

$c^2 = a^2 + b^2$

$c^2 = 49 + 4$

$c^2 = 53$

Center: $(h, k) = (0, 0)$

Vertices: $(h \pm a, k) = (0 \pm 7, 0) = (\pm 7, 0)$

Foci: $(h \pm c, k) = (0 \pm \sqrt{53}, 0) = (\pm \sqrt{53}, 0)$

Graph:

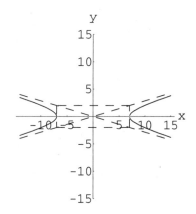

(Hyperbolas)

24. $\dfrac{(y-1)^2}{4} - \dfrac{(x+2)^2}{1} = 1$

The hyperbola opens up and down.

$(h, k) = (-2, 1)$, $a^2 = 1$, $b^2 = 4$, $c^2 = 1 + 4 = 5$

Center: $(h, k) = (-2, 1)$

Vertices: $(h, k \pm b) = (-2, 1 \pm 2) = (-2, 3)$ and $(-2, -1)$

Foci: $(h, k \pm c) = (-2, 1 \pm \sqrt{5})$

Graph:

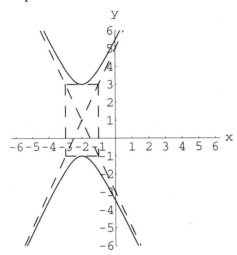

(Hyperbolas)

25. $3x^2 - 2y^2 - 6x - 8y - 11 = 0$

$$3x^2 - 6x - 2y^2 - 8y = 11 \qquad \text{Group } x^2 \text{ and } x, y^2 \text{ and } y \text{ terms.}$$

$$3(x^2 - 2x) - 2(y^2 + 4y) = 11 \qquad \text{Factor out coefficients of the squared terms.}$$

$$3(x^2 - 2x + 1) - 2(y^2 + 4y + 4) = 11 + 3(1) + (-2)(4)$$

$$3(x - 1)^2 - 2(y + 2)^2 = 6$$

$$\frac{3(x - 1)^2}{6} - \frac{2(y + 2)^2}{6} = \frac{6}{6} \qquad \text{Divide each term by 6.}$$

$$\frac{(x - 1)^2}{2} - \frac{(y + 2)^2}{3} = 1 \qquad \text{Write in standard form.}$$

The hyperbola opens left and right.

$(h, k) = (1, -2)$, $a^2 = 2$ so $a = \sqrt{2}$, $b^2 = 3$ so $b = \sqrt{3}$.

$c^2 = a^2 + b^2 = 2 + 3 = 5$ so $c = \sqrt{5}$.

Center: $(h, k) = (1, -2)$

Vertices: $(h \pm a, k) = (1 \pm \sqrt{2}, -2)$

Foci: $(h \pm c, k) = (1 \pm \sqrt{5}, -2)$

Graph:

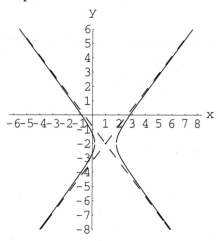

(Hyperbolas)

26.   Center at the origin, vertices at $(\pm 2, 0)$, and foci at $(\pm 4, 0)$. A quick sketch reveals that the hyperbola opens left and right.

$(h, k) = (0, 0)$ because the center is at the origin.

$a = 2$ so $a^2 = 4$

$c = 4$ so $c^2 = 16$

$c^2 = a^2 + b^2$

$16 = 4 + b^2$

$12 = b^2$

The standard form is:

$$\frac{(x-0)^2}{4} - \frac{(y-0)^2}{12} = 1 \text{ or } \frac{x^2}{4} - \frac{y^2}{12} = 1$$

(Hyperbolas)

27.   Center at the origin, vertices at $(0, \pm 3)$ and asymptotes $y = \pm 3x$.

Sketch the given information:

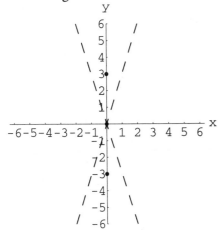

$(h, k) = (0, 0)$. $b = 3$ so $b^2 = 9$.

The formula for the asymptotes is:

$$y - k = \pm \frac{b}{a}(x - h)$$

$$y - 0 = \pm \frac{b}{a}(x - 0) \qquad \text{Use } (h, k) = (0, 0).$$

$$y = \pm \frac{b}{a}x$$

Therefore,

$$\frac{b}{a} = \frac{3}{1}$$

$$\frac{3}{a} = \frac{3}{1} \qquad \text{We know } b = 3.$$

$$a = 1 \qquad \text{Solve for } a.$$

Thus, the equation of the hyperbola is:

$$\frac{y^2}{9} - \frac{x^2}{1} = 1$$

(Hyperbolas)

28.   Vertices at $(2, 0)$ and $(6, 0)$ and foci at $(-1, 0)$ and $(9, 0)$.

Sketch the given information:

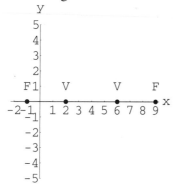

The center must be halfway between the vertices, so $(h, k) = (4, 0)$.

$a = 2$ so $a^2 = 4$, $c = 5$ so $c^2 = 25$.

$c^2 = a^2 + b^2$

$25 = 4 + b^2$

$21 = b^2$

The standard form is:

$$\frac{(x-4)^2}{4} - \frac{y^2}{21} = 1$$

(Hyperbolas)

29.   Vertices at $(3, 2)$ and $(-3, 2)$ and passing through $(5, 0)$.

Sketch the given information:

The hyperbola opens left and right and has center at $(h, k) = (0, 2)$.

$a = 3$ so $a^2 = 9$. Using the values we know, we have

$$\frac{x^2}{9} - \frac{(y-2)^2}{b^2} = 1$$

Since the hyperbola passes through the point $(5, 0)$, we can also substitute the x and y values into the equation to find b:

$$\frac{5^2}{9} - \frac{(0-2)^2}{b^2} = 1$$

$$\frac{25}{9} - \frac{4}{b^2} = 1$$

$$9b^2 \cdot \frac{25}{9} - 9b^2 \cdot \frac{4}{b^2} = 9b^2(1) \qquad \text{Multiply each term by } 9b^2.$$

$$25b^2 - 36 = 9b^2$$

$$16b^2 = 36$$

$$b^2 = \frac{36}{16} = \frac{9}{4}$$

Therefore, the standard form is:

$$\frac{x^2}{9} - \frac{(y-2)^2}{9/4} = 1$$

(Hyperbolas)

30. $4y^2 + 2x - 2y + 8 = 0$

Since $A = 0$, the graph is a parabola.

(General second-degree equation in two variables)

31. $6x^2 + 6y^2 - 2x + 3y - 6 = 0$

Since $A = C = 6$, the graph is a circle.

(General second-degree equation in two variables)

32. $2x^2 - 4y^2 + 4y - 10 = 0$

Since the signs of A and C are different $(A = 2, C = -4)$, the graph is a hyperbola.

(General second-degree equation in two variables)

33. $-3x^2 - 3y^2 + 2x - 4y + 8 = 0$

Since $A = C = -3$, the graph is a circle.

(General second-degree equation in two variables)

34. $-2x^2 - 4y^2 + 8x + 12 = 0$

Since the signs of A and C are the same $(A = -2, C = -4)$, the graph is an ellipse.

(General second-degree equation in two variables)

Grade Yourself

Circle the question numbers that you had incorrect. Then indicate the number of questions you missed. If you answered more than three questions incorrectly, you will have to focus on that topic. If a topic has fewer than three questions and you had at least one wrong, we suggest you study that topic. Read your textbook or a review book or ask your teacher for help.

Subject: Conic Sections

Topic	Question Numbers	Number Incorrect
Parabolas	1, 2, 3, 4, 5, 6, 7, 8, 9, 10, 11, 12	
Ellipses	13, 14, 15, 16, 17, 18, 19, 20	
Hyperbolas	21, 22, 23, 24, 25, 26, 27, 28, 29	
General second-degree equation in two variables	30, 31, 32, 33, 34	

Systems of Equations, Matrices, and Systems of Inequalities

8

Brief Yourself

This chapter includes questions about solving systems of equations algebraically, graphically, and by using Gaussian elimination, matrices, and Cramer's Rule. There are also questions about matrix operations, including finding the inverse of a square matrix. Questions about systems of inequalities and linear programming are also contained in this chapter.

Substitution and elimination are the techniques generally used to solve systems of two equations in two variables. The solution(s) to a system corresponds to the intersection point(s) of the graphs of the system. Note that there may be no solutions, 1 solution, or more than 1 solution. If the test directions do not specify which method to use and one equation is already solved for one variable, use substitution (it's usually more convenient). Note that elimination can be used only when both equations contain the same variable raised to the same power. It is easy to make errors when working systems of equation problems, so check all answers as time allows.

A linear system may be solved by Gaussian elimination, where the elementary row operations are performed (interchange any two equations, multiply both sides of any equation by the same nonzero number, and add a multiple of any equation to any other equation) to write the system in triangular form. The solution can be found by back-substitution where the last equation is solved first, and that solution is used to solve the previous equation, etc., until all equations have been solved. If, in the process, all variables in an equation are eliminated and a false statement is produced (such as $0 = -2$), the system is inconsistent and has an empty solution set. If, in the process, all variables in an equation are eliminated and a true statement is produced (such as $0 = 0$), the system has an infinite number of solutions. In this case, the answer is generally written using a parameter, such as t.

Systems of linear equations can also be solved by writing the system as an augmented matrix (strip off the letters) and using the elementary row operations to change the matrix to row echelon form. In a linear system of three equations in three unknowns, try to produce 0's as shown:

$$\begin{bmatrix} a & b & c & d \\ 0 & e & f & g \\ 0 & 0 & h & i \end{bmatrix}$$

From here, the system can be solved by back-substitution.

The usual operations of addition, subtraction, scalar multiplication, and multiplication can be performed on matrices, providing the order of matrices is appropriate (for example, only matrices of the same order can be added or subtracted). Two matrices are equal if they have the same order and their entries are equal. The elements in a matrix are located by their row and column position, so that element a_{13} is in the 1st row, 3rd column of matrix A.

Division of matrices is done with the multiplicative inverse (usually called the inverse). The inverse of a square matrix (one that has an equal number of rows and columns) is found by adjoining the identity matrix to the given matrix and using elementary row operations to turn the given matrix into the identity. The adjoined matrix is, at that point, the inverse. The inverse of a 2×2 matrix can be found by use of the formula:

$$\text{If } A = \begin{bmatrix} a & b \\ c & d \end{bmatrix}, A^{-1} = \frac{1}{ad - bc} \begin{bmatrix} d & -b \\ -c & a \end{bmatrix}.$$

A system of linear equations written in matrix form $Ax = B$ can be solved by multiplying A^{-1} times B.

The determinant of the 2×2 matrix $\begin{bmatrix} a & b \\ c & d \end{bmatrix}$ is written $\begin{vmatrix} a & b \\ c & d \end{vmatrix}$ and is equal to $ad - bc$.

The determinant of $\begin{bmatrix} a & b & c \\ d & e & f \\ g & h & i \end{bmatrix}$ can be found by:

1. Writing the first two columns adjacent to the original three columns.
2. Multiplying the elements down the diagonals containing three entries and then adding.
3. Multiplying the elements up the diagonals containing three entries and then adding.
4. Subtracting the answer in step 3 from the answer in step 2.

Note: The method stated above is valid only for 3×3 matrices.

Determinants can also be found by expanding along any row or column. Multiply each element in the row or column by its cofactor and sum the products. Note that the cofactor is the product of:

$(-1)^{\text{row number + column number of the element}}$ times the determinant formed when the row and column containing the element are deleted.

Cramer's Rule uses determinants to solve systems of equations. The rule is stated below for solving two equations in two variables and three equations in three variables.

Note: The denominator for x and y (and for x, y, and z in the three-equation case) is the same. Thus, once you have found that value, you can use it for each denominator. Also, Cramer's Rule is *not* valid if the denominator equals 0.

The solution to $\begin{array}{l} a_1x + b_1y = c \\ a_2x + b_2y = d \end{array}$ is $x = \dfrac{\begin{vmatrix} c & b_1 \\ d & b_2 \end{vmatrix}}{\begin{vmatrix} a_1 & b_1 \\ a_2 & b_2 \end{vmatrix}}$ and $y = \dfrac{\begin{vmatrix} a_1 & c \\ a_2 & d \end{vmatrix}}{\begin{vmatrix} a_1 & b_1 \\ a_2 & b_2 \end{vmatrix}}.$

The solution to $\begin{array}{l} a_1x + b_1y + c_1z = d \\ a_2x + b_2y + c_2z = e \\ a_3x + b_3y + c_3z = f \end{array}$ is $x = \dfrac{\begin{vmatrix} d & b_1 & c_1 \\ e & b_2 & c_2 \\ f & b_3 & c_3 \end{vmatrix}}{\begin{vmatrix} a_1 & b_1 & c_1 \\ a_2 & b_2 & c_2 \\ a_3 & b_3 & c_3 \end{vmatrix}}$ $y = \dfrac{\begin{vmatrix} a_1 & d & c_1 \\ a_2 & e & c_2 \\ a_3 & f & c_3 \end{vmatrix}}{\begin{vmatrix} a_1 & b_1 & c_1 \\ a_2 & b_2 & c_2 \\ a_3 & b_3 & c_3 \end{vmatrix}}$

$z = \dfrac{\begin{vmatrix} a_1 & b_1 & d \\ a_2 & b_2 & e \\ a_3 & b_3 & f \end{vmatrix}}{\begin{vmatrix} a_1 & b_1 & c_1 \\ a_2 & b_2 & c_2 \\ a_3 & b_3 & c_3 \end{vmatrix}}.$

Systems of inequalities are solved by graphing. Graph each inequality and shade the appropriate region (choose test points to determine which regions should be shaded). Darken the area where the shadings overlap for your final answer. Remember that boundary curves are solid if the inequality contains ≤ or ≥ and are dotted if the inequality contains < or >. These techniques are used to solve linear programming problems where a maximum or minimum value of a given objective function is desired. Graph and shade as before, but test each vertex (points of intersection of the lines) in the objective function to determine which vertex yields the optimum value.

Test Yourself

Find the solution set.

1. $3x - 4y = 1$

   $2x + 6y = 18$

2. $2x + 4y = 8$

   $3x - 5y = -21$

3. $6x - 3y = -15$

   $y = x + 4$

4. $\dfrac{1}{4}x - \dfrac{1}{2}y = -\dfrac{5}{4}$

   $\dfrac{1}{3}x + \dfrac{1}{4}y = -\dfrac{3}{4}$

5. $x^2 + y^2 = 4$

    $2x - 3y = 4$

6. $\dfrac{2}{x} - \dfrac{5}{y} = -\dfrac{1}{6}$

    $\dfrac{3}{x} + \dfrac{4}{y} = \dfrac{31}{60}$

7. Use back-substitution to solve:

    $6x + 4y + 4z = 2$

    $y - 11z = 35$

    $7z = -21$

8. Use back-substitution to solve:

    $5x - y + 3z + w = 0$

    $3y - 2z - w = 9$

    $z + w = -3$

    $8w = -24$

Solve each system.

9. $2x + 3y - z = -8$

    $x - 3y + 2z = 7$

    $5x + 60y + 2z = 20$

10. $2x + y + 3z = 4$

    $2x + 3y - z = -4$

    $3x - 2y - 5z = -13$

11. $x - 3y = -14$

    $2x + y + 5z = 0$

    $-2y + 7z = -8$

12. $3x - 2y = 0$

    $2x - 3y + z = 0$

    $3x - 7y + 3z = 0$

13. Form the augmented matrix for the system.

    $3x - 6y + z = -5$

    $x - 5y - 3z = -5$

    $-x - y + 4z = 6$

14. Write the system of linear equations represented by the augmented matrix (use variables x_1, x_2, x_3).

$$\begin{bmatrix} 2 & 0 & 1 & 3 \\ 0 & 1 & -2 & 6 \\ 4 & 5 & 1 & 8 \end{bmatrix}$$

Find the solution set using an augmented matrix.

15. $3x - 4y = 1$

    $2x + 6y = 18$

16. $x - 2y + 5z = 0$

    $2x + 2y + z = 3$

    $3x - 4y - 3z = 1$

Use the following matrices to perform the indicated operations.

$$A = \begin{bmatrix} 2 & -1 & 3 \\ 4 & 0 & 5 \\ 2 & 1 & 2 \end{bmatrix} \quad B = \begin{bmatrix} 1 & 2 \\ 3 & -2 \\ 4 & 1 \end{bmatrix}$$

$$C = \begin{bmatrix} 2 & 5 \\ -1 & 1 \end{bmatrix} \quad D = \begin{bmatrix} -1 & -3 \\ 0 & 1 \\ 6 & 5 \end{bmatrix}$$

17. $B + D$

18. $D - B$

19. AB

20. $-4C$

21. C^2

22. Find C^{-1}.

23. Show that E is the inverse of F if $E = \begin{bmatrix} 3 & 1 \\ 0 & -1 \end{bmatrix}$

    and $F = \begin{bmatrix} \dfrac{1}{3} & \dfrac{1}{3} \\ 0 & -1 \end{bmatrix}$.

24. Find the inverse of

$$\begin{bmatrix} -2 & 0 & -1 \\ \frac{1}{2} & 1 & -1 \\ 4 & 0 & 1 \end{bmatrix}$$

25. Solve the system using the inverse matrix found in problem 22.

$$2x + 5y = -17$$

$$-x + y = -9$$

26. Solve the system using the inverse matrix found in problem 24.

$$-2x - z = -1$$

$$\frac{1}{2}x + y - 2 = -6$$

$$4x + z = -3$$

Find the value of each determinant.

27. $\begin{vmatrix} 2 & 5 \\ 1 & 6 \end{vmatrix}$

28. $\begin{vmatrix} 4 & 2 \\ 3 & 0 \end{vmatrix}$

29. $\begin{vmatrix} 3 & 2 & 2 \\ -1 & -1 & 1 \\ 0 & 1 & 2 \end{vmatrix}$

30. $\begin{vmatrix} -1 & 6 & 3 \\ 2 & -4 & 1 \\ -3 & -2 & 5 \end{vmatrix}$

31. Use Cramer's Rule to solve the system.

$$2x + 3y = 5$$

$$-3x - 5y = -9$$

32. Solve for x using Cramer's Rule. Set up the determinants needed to solve for y and z.

$$2x + y + 3z = 4$$

$$2x + 3y - z = -4$$

$$3x - 2y - 5z = -13$$

Graph the solution set of each system of inequalities.

33. $x + y \le 2$

$-x + y \le 2$

$x \ge 0$

34. $x^2 + y^2 \le 16$

$x^2 + y^2 \ge 4$

35. $y \ge x^2$

$y < \frac{1}{2}x + 2$

36. Find the minimum and maximum value for the given objective function subject to the given constraints.

Objective function: $C = 6x + 9y$

Constraints: $x \ge 0$

$y \ge 0$

$4x + 3y \le 24$

$x + 2y \ge 6$

37. A merchant plans to sell two models of hard drive which cost $100 and $300. The $100 model yields a profit of $100 and the $300 model yields a profit of $200. The merchant estimates that the total monthly demand will not exceed 100 units. The merchant does not want to invest more than $20,000 in inventory for these products. Find the number of units of each model that should be stocked in order to maximize profit.

Check Yourself

1. $3x - 4y = 1$ This system can be solved by elimination.
      $2x + 6y = 18$

      $2(3x - 4y = 1)$ Multiply the first equation by 2.
      $-3(2x + 6y = 18)$ Multiply the second equation by –3.

      $6x - 8y = 2$ Add the equations.
      $-6x - 18y = -54$

      $-26y = -52$

      $y = 2$ Solve for y.

      $3x - 4(2) = 1$ Substitute $y = 2$ into the first equation.

      $3x - 8 = 1$ Solve for x.

      $3x = 9$

      $x = 3$

    $(3, 2)$ Write the solution as an ordered pair.

    (Systems of equations)

2. $2x + 4y = 8$ Solve by elimination.
      $3x - 5y = -21$

      $5(2x + 4y = 8)$ Eliminate y.
      $4(3x - 5y = -21)$

      $10x + 20y = 40$
      $12x - 20y = -84$

      $22x = -44$ Add the equations.

      $x = -2$ Solve for x.

      $2(-2) + 4y = 8$ Substitute $x = -2$ into the first equation.

      $-4 + 4y = 8$ Solve for y.

      $4y = 12$

      $y = 3$

    $(-2, 3)$ Write the solution as an ordered pair.

    (Systems of equations)

3.
$$6x - 3y = -15$$
$$y = x + 4$$

Solve by substitution.

$$6x - 3(x + 4) = -15$$
$$6x - 3x - 12 = -15$$

Substitute $(x + 4)$ for y in the first equation.
Multiply.

$$3x = -3$$

Solve for x.

$$x = -1$$

$$y = (-1) + 4$$

Substitute $x = -1$ into the second equation.

$$y = 3$$

$(-1, 3)$

Write the solution as an ordered pair.

(Systems of equations)

4.
$$\frac{1}{4}x - \frac{1}{2}y = -\frac{5}{4}$$
$$\frac{1}{3}x + \frac{1}{4}y = -\frac{3}{4}$$

$$4\left(\frac{1}{4}x - \frac{1}{2}y = -\frac{5}{4}\right)$$

Multiply to eliminate the fractions.

$$12\left(\frac{1}{3}x + \frac{1}{4}y = -\frac{3}{4}\right)$$

$$x - 2y = -5$$
$$4x + 3y = -9$$

Now use elimination to solve the system.

$$3(x - 2y = -5)$$
$$2(4x + 3y = -9)$$

Multiply the first equation by 3.
Multiply the second equation by 2.

$$3x - 6y = -15$$
$$8x + 6y = -18$$

$$11x = -33$$

Add the equations.

$$x = -3$$

Solve for x.

While you could substitute the x-value into one of the original equations, it will be easier to substitute into one of the equivalent equations without the fractions.

$$(-3) - 2y = -5$$

$$-2y = -2$$

$$y = 1$$

$(-3, 1)$

Write the solution as an ordered pair.

(Systems of equations)

5. $x^2 + y^2 = 4$

    $2x - 3y = 4$

Note that elimination cannot be used in this case because adding multiples of the equations will not cause a variable to drop out. Use substitution, solving the second equation for x:

$$2x - 3y = 4$$

$$2x = 3y + 4 \qquad\qquad \text{Add } 3y \text{ to both sides.}$$

$$x = \frac{3}{2}y + 2 \qquad\qquad \text{Divide each term by 2.}$$

$$\left(\frac{3}{2}y + 2\right)^2 + y^2 = 4 \qquad\qquad \text{Substitute for } x \text{ in the first equation.}$$

$$\frac{9}{4}y^2 + 6y + 4 + y^2 = 4 \qquad\qquad \text{Multiply.}$$

$$9y^2 + 24y + 16 + 4y^2 = 16 \qquad\qquad \text{Clear fractions.}$$

$$13y^2 + 24y = 0 \qquad\qquad \text{Combine similar terms.}$$

$$y(13y + 24) = 0 \qquad\qquad \text{Factor.}$$

$$y = 0 \text{ or } y = -\frac{24}{13} \qquad\qquad \text{Set each factor equal to 0 and solve.}$$

When $y = 0$,

$$x = \frac{3}{2}(0) + 2$$

$$x = 2$$

When $y = -\frac{24}{13}$

$$x = \frac{3}{2}\left(-\frac{24}{13}\right) + 2$$

$$x = -\frac{10}{13}$$

$(2, 0)$ and $(-\frac{10}{13}, -\frac{24}{13})$ $\qquad\qquad$ Write the solutions as ordered pairs.

(Systems of equations)

6. $\dfrac{2}{x} - \dfrac{5}{y} = -\dfrac{1}{6}$

    $\dfrac{3}{x} + \dfrac{4}{y} = \dfrac{31}{60}$

Although fractions are generally cleared, in this case clearing leads to a system that is more difficult to solve. However, the y term can be eliminated by multiplying the first equation by 4 and the second equation by 5:

$$\frac{8}{x} - \frac{20}{y} = -\frac{4}{6}$$

$$\frac{15}{x} + \frac{20}{y} = \frac{31}{12}$$

$$\frac{23}{x} = \frac{23}{12} \qquad\qquad -\frac{4}{6} + \frac{31}{12} = -\frac{8}{12} + \frac{31}{12} = \frac{23}{12}$$

$$x = 12 \qquad\qquad \text{Solve for } x.$$

$$\frac{2}{12} - \frac{5}{y} = -\frac{1}{6} \qquad\qquad \text{Substitute } x = 12 \text{ into the first equation.}$$

$$12y\left(\frac{2}{12} - \frac{5}{y} = -\frac{1}{6}\right) \qquad\qquad \text{Clear fractions.}$$

$$2y - 60 = -2y \qquad\qquad \text{Solve for } y.$$

$$4y = 60$$

$$y = 15$$

$$(12, 15) \qquad\qquad \text{Write the solution as an ordered pair.}$$

(Systems of equations)

7. $\quad 6x + 4y + 4z = 2$

$\qquad\qquad y - 11z = 35$

$\qquad\qquad\qquad 7z = -21$

Since $7z = -21, z = -3.$

$$y - 11(-3) = 35 \qquad\qquad \text{Substitute } z = -3 \text{ into the second equation.}$$

$$y = 2 \qquad\qquad \text{Solve for } y.$$

$$6x + 4(2) + 4(-3) = 2 \qquad\qquad \text{Substitute into the first equation.}$$

$$6x + 8 - 12 = 2 \qquad\qquad \text{Solve for } x.$$

$$6x = 6$$

$$x = 1$$

$$(1, 2, -3) \qquad\qquad \text{Write the solution as an ordered triple.}$$

(Linear systems)

8. $5x - y + 3z + w = 0$
$3y - 2z - w = 9$
$z + w = -3$
$8w = -24$

Start with the last equation and work up.

$$8w = -24$$

$$w = -3$$

$$z + (-3) = -3$$

$$z = 0$$

$$3y - 2(0) - (-3) = 9$$

$$y = 2$$

$$5x - (2) + 3(0) + (-3) = 0$$

$$x = 1$$

$(1, 2, 0, -3)$ Write the solution as an ordered 4-tuple.

(Linear systems)

9.   (1) $2x + 3y - z = -8$
(2) $x - 3y + 2z = 7$
(3) $5x + 60y + 2z = 20$

Eliminate x using equations (1) and (2):

Current System   **Multiply equation (2) by –2 and add to equation (1):**

(1) $2x + 3y - z = -8$ (1) $2x + 3y - z = -8$
(2) $x - 3y + 2z = 7$ –2(2) $-2x + 6y - 4z = -14$
(3) $5x + 60y + 2z = 20$ $9y - 5z = -22$ Add (1) and –2(2).

Replace equation (2) with $9y - 5z = -22$. Eliminate x using equations (1) and (3):

Current System   **Multiply equation (1) by –5 and add to 2 times equation (3):**

(1) $2x + 3y - z = -8$ –5(1) $-10x - 15y + 5z = 40$
(2) $9y - 5z = -22$ 2(3) $10x + 120y + 4z = 40$
(3) $5x + 60y + 2z = 20$ $105y + 9z = 80$ Add –5(1) and 2(3).

Replace equation (3) with $105y + 9z = 80$. Now eliminate y using equations (2) and (3):

Current System:

Multiply equation (2) by –105 and add to 9 times equation (3):

(1)	$2x + 3y - z = -8$	$-105(2)$	$-945y + 525z = 2310$	
(2)	$9y - 5z = -22$	$9(3)$	$945y + 81z = 720$	
(3)	$105y + 9z = 80$		$606z = 3030$	Add $-105(2)$ and $9(3)$.

Replace equation (3) with $606z = 3030$:

(1) $2x + 3y - z = -8$
(2) $9y - 5z = -22$
(3) $606z = 3030$

This system can now be solved by back-substitution:

$$606z = 3030$$

$$z = 5$$

$$9y - 5(5) = -22$$

$$9y - 25 = -22$$

$$9y = 3$$

$$y = \frac{1}{3}$$

$$2x + 3\left(\frac{1}{3}\right) - (5) = -8$$

$$2x - 4 = -8$$

$$2x = -4$$

$$x = -2$$

$(-2, \frac{1}{3}, 5)$ is the solution.

(Linear systems)

10. (1) $2x + y + 3z = 4$
       (2) $2x + 3y - z = -4$
       (3) $3x - 2y - 5z = -13$

Eliminate x using equations (1) and (2). Multiply -1 times equation (1) and add equation (2):

$$-1(1)- 2x - y - 3z = -4$$
$$(2) \quad 2x + 3y - z = -4$$
$$2y - 4z = -8$$

Replace equation (2) with $2y - 4z = -8$, so the new system becomes:

(1) $\quad 2x + y + 3z = 4$
(2) $\qquad 2y - 4z = -8$
(3) $\quad 3x - 2y - 5z = -13$

Eliminate x using equations (1) and (3). Multiply -3 times equation (1) and add 2 times equation (3):

$$-3(1)- 6x - 3y - 9z = -12$$
$$2(3) \quad 6x - 4y - 10z = -26$$
$$-7y - 19z = -38$$

Replace equation (3) with $-7y - 19z = -38$, so the new system becomes:

(1) $\quad 2x + y + 3z = 4$
(2) $\qquad 2y - 4z = -8$
(3) $\qquad -7y - 19z = -38$

Now eliminate y from equation (3) using equations (2) and (3). Multiply 7 times equation (2) and add 2 times equation (3):

$$7(2) \quad 14y - 28z = -56$$
$$2(3) \quad -14y - 38z = -76$$
$$-66z = -132$$

Replace eqaution (3) with $-66z = -132$, so the new system becomes:

(1) $\quad 2x + y + 3z = 4$
(2) $\qquad 2y - 4z = -8$
(3) $\qquad -66z = -132$

Now solve by back-substitution:

$-66z = -132$, so $\quad z = 2$

$2y - 4(2) = -8$, so $y = 0$

$2x + (0) + 3(2) = 4$, so $x = -1$

$(-1, 0, 2)$ is the solution.

(Linear systems)

11. (1) $x - 3y = -14$
     (2) $2x + y + 5z = 0$
     (3) $-2y + 7z = -8$

Eliminate x from equation (2). It is already missing in equation (3).

(1) $x - 3y = -14$ $-2(1)$ $-2x + 6y = 28$
(2) $7y + 5z = 28$ (2) $2x + y + 5z = 0$
(3) $-2y + 7z = -8$ $7y + 5z = 28$

Now eliminate y from equation (3):

(1) $x - 3y = -14$ $2(2)$ $14y + 10z = 56$
(2) $7y + 5z = 28$ $7(3)$ $-14y + 49z = -56$
(3) $59z = 0$ $59z = 0$

Now solve by back-substitution:

$$59z = 0$$

$$z = 0$$

$$7y + 5(0) = 28$$

$$y = 4$$

$$x - 3(4) = -14$$

$$x = -2$$

$(-2, 4, 0)$ is the solution.

(Linear systems)

12. (1) $3x - 2y \quad = 0$
     (2) $2x - 3y + z = 0$
     (3) $3x - 7y + 3z = 0$

Eliminate x from equations (2) and (3):

(1) $3x - 2y = 0$ $-2(1)$ $-6x + 4y = 0$
(2) $-2y + 3z = 0$ $3(2)$ $6x - 9y + 3z = 0$
(3) $3x - 7y + 3z = 0$ $-5y + 3z = 0$

(1) $3x - 2y = 0$ $-1(1)$ $-3x + 2y = 0$
(2) $-5y + 3z = 0$ (3) $3x - 7y + 3z = 0$
(3) $-5y + 3z = 0$ $-5y + 3z = 0$

Now eliminate y from equation (3):

(1) $3x - 2y = 0$

(2) $-5y + 3z = 0$

(3) $0 = 0$

$-1(2)$ $5y - 3z = 0$

(3) $-5y + 3z = 0$

$0 = 0$

Notice that all the variables were eliminated in equation (3) and a true statement resulted ($0 = 0$). Introduce a parameter, say t, and let $z = t$.

Now solve by back-substitution, using t for z:

$$-5y + 3(t) = 0$$

$$-5y = -3t$$

$$y = \frac{3}{5}t$$

$$3x - 2\left(\frac{3}{5}t\right) = 0$$

$$3x = \frac{6}{5}t$$

$$x = \frac{2}{5}t$$

$\left(\frac{2}{5}t, \frac{3}{5}t, t\right)$ is the solution for t a real number.

(Linear systems)

13.
$$\begin{bmatrix} 3 & -6 & 1 & \vdots & -5 \\ 1 & -5 & -3 & \vdots & -5 \\ -1 & -1 & 4 & \vdots & 6 \end{bmatrix}$$

(Augmented matrix)

14. $2x_1 + 0x_2 + x_3 = 3$

$0x_1 + x_2 - 2x_3 = 6$

$4x_1 + 5x_2 + x_3 = 8$

(Augmented matrix)

15. $3x - 4y = 1$

$2x + 6y = 18$

First write the augmented matrix:

$$\begin{bmatrix} 3 & -4 & | & 1 \\ 2 & 6 & | & 18 \end{bmatrix}$$

Use row operations to produce row echelon form:

$$\begin{bmatrix} 3 & -4 & | & 1 \\ 0 & 26 & | & 52 \end{bmatrix}$$

$$\begin{array}{r} -2 \cdot R_1 \quad -6 \quad 8 \quad -2 \\ 3 \cdot R_2 \quad \underline{6 \quad 18 \quad 54} \\ 0 \quad 26 \quad 52 \end{array}$$

You can use back-substitution to complete the solution:

$26y = 52$ so $y = 2$.

$3x - 4(2) = 1$

$3x = 9$

$x = 3$

$(3, 2)$ is the solution.

(Augmented matrix)

16. $x - 2y + 5z = 0$

$2x + 2y + z = 3$

$3x - 4y - 3z = 1$

$$\begin{bmatrix} 1 & -2 & 5 & | & 0 \\ 2 & 2 & 1 & | & 3 \\ 3 & -4 & -3 & | & 1 \end{bmatrix}$$

Write the augmented matrix.

$$\begin{bmatrix} 1 & -2 & 5 & | & 0 \\ 0 & 6 & -9 & | & 3 \\ 3 & -4 & -3 & | & 1 \end{bmatrix}$$

$$\begin{array}{r} -2 \cdot R_1 \quad -2 \quad 4 \quad -10 \quad 0 \\ R_2 \quad \underline{2 \quad 2 \quad 1 \quad 3} \\ 0 \quad 6 \quad -9 \quad 3 \end{array}$$

$$\begin{bmatrix} 1 & -2 & 5 & | & 0 \\ 0 & 6 & -9 & | & 3 \\ 0 & 2 & -18 & | & 1 \end{bmatrix}$$

$$\begin{array}{r} -3 \cdot R_1 \quad -3 \quad 6 \quad -15 \quad 0 \\ R_3 \quad \underline{3 \quad -4 \quad -3 \quad 1} \\ 0 \quad 2 \quad -18 \quad 1 \end{array}$$

$$\begin{bmatrix} 1 & -2 & 5 & | & 0 \\ 0 & 6 & -9 & | & 3 \\ 0 & 0 & 45 & | & 0 \end{bmatrix}$$

$$\begin{array}{r} R_2 \quad 0 \quad 6 \quad -9 \quad 3 \\ -3R_3 \quad \underline{0 \quad -6 \quad 54 \quad -3} \\ 0 \quad 0 \quad 45 \quad 0 \end{array}$$

Note that the process could be continued to produce reduced row echelon form, but it is probably more convenient to solve from here by back-substitution.

$$45z = 0$$

$$z = 0$$

$$6y - 9(0) = 3$$

$$y = \frac{1}{2}$$

$$x - 2\left(\frac{1}{2}\right) + 5(0) = 0$$

$$x = 1$$

$(1, \dfrac{1}{2}, 0)$ is the solution.

(Augmented matrix)

17. $\begin{bmatrix} 1 & 2 \\ 3 & -2 \\ 4 & 1 \end{bmatrix} + \begin{bmatrix} -1 & -3 \\ 0 & 1 \\ 6 & 5 \end{bmatrix} = \begin{bmatrix} 0 & -1 \\ 3 & -1 \\ 10 & 6 \end{bmatrix}$

Add the corresponding entries.

(Matrix algebra)

18. $\begin{bmatrix} -1 & -3 \\ 0 & 1 \\ 6 & 5 \end{bmatrix} - \begin{bmatrix} 1 & 2 \\ 3 & -2 \\ 4 & 1 \end{bmatrix} = \begin{bmatrix} -2 & -5 \\ -3 & 3 \\ 2 & 4 \end{bmatrix}$

Subtract the corresponding entries.

(Matrix algebra)

19. $\begin{bmatrix} 2 & -1 & 3 \\ 4 & 0 & 5 \\ 2 & 1 & 2 \end{bmatrix} \begin{bmatrix} 1 & 2 \\ 3 & -2 \\ 4 & 1 \end{bmatrix} = \begin{bmatrix} 2-3+12 & 4+2+3 \\ 4+0+20 & 8+0+5 \\ 2+3+8 & 4-2+2 \end{bmatrix} = \begin{bmatrix} 11 & 9 \\ 24 & 13 \\ 13 & 4 \end{bmatrix}$

(Matrix algebra)

20. $-4C = -4\begin{bmatrix} 2 & 5 \\ -1 & 1 \end{bmatrix} = \begin{bmatrix} -8 & -20 \\ 4 & -4 \end{bmatrix}$

(Matrix algebra)

21. $C^2 = C \cdot C = \begin{bmatrix} 2 & 5 \\ -1 & 1 \end{bmatrix}\begin{bmatrix} 2 & 5 \\ -1 & 1 \end{bmatrix} = \begin{bmatrix} -1 & 15 \\ -3 & -4 \end{bmatrix}$

(Matrix algebra)

22. $C = \begin{bmatrix} 2 & 5 \\ -1 & 1 \end{bmatrix}$

$C^{-1} = \dfrac{1}{2-(-5)}\begin{bmatrix} 1 & -5 \\ 1 & 2 \end{bmatrix} = \dfrac{1}{7}\begin{bmatrix} 1 & -5 \\ 1 & 2 \end{bmatrix} = \begin{bmatrix} \dfrac{1}{7} & -\dfrac{5}{7} \\ \dfrac{1}{7} & \dfrac{2}{7} \end{bmatrix}$

(Inverse of a square matrix)

23. To show that E is the inverse of F, show that $E \cdot F = I$ where I is the identity matrix.

$\begin{bmatrix} 3 & 1 \\ 0 & -1 \end{bmatrix}\begin{bmatrix} \dfrac{1}{3} & \dfrac{1}{3} \\ 0 & -1 \end{bmatrix} = \begin{bmatrix} 1+0 & 1-1 \\ 0+0 & 0+1 \end{bmatrix} = \begin{bmatrix} 1 & 0 \\ 0 & 1 \end{bmatrix}$

(Inverse of a square matrix)

24. $\left[\begin{array}{ccc|ccc} -2 & 0 & -1 & 1 & 0 & 0 \\ \dfrac{1}{2} & 1 & -1 & 0 & 1 & 0 \\ 4 & 0 & 1 & 0 & 0 & 1 \end{array}\right]$ Adjoin the identity matrix.

$\left[\begin{array}{ccc|ccc} -2 & 0 & -1 & 1 & 0 & 0 \\ 0 & 4 & -5 & 1 & 4 & 0 \\ 0 & 0 & -1 & 2 & 0 & 1 \end{array}\right]$

$$\begin{array}{rrrrrrr} R_1 & -2 & 0 & -1 & 1 & 0 & 0 \\ 4 \cdot R_2 & 2 & 4 & -4 & 0 & 4 & 0 \\ \hline & 0 & 4 & -5 & 1 & 4 & 0 \end{array}$$

$$\begin{array}{rrrrrrr} 2 \cdot R_1 & -4 & 0 & -2 & 2 & 0 & 0 \\ R_3 & 4 & 0 & 1 & 0 & 0 & 1 \\ \hline & 0 & 0 & -1 & 2 & 0 & 1 \end{array}$$

$\left[\begin{array}{ccc|ccc} -2 & 0 & -1 & 1 & 0 & 0 \\ 0 & 4 & 0 & -9 & 4 & -5 \\ 0 & 0 & -1 & 2 & 0 & 1 \end{array}\right]$

$$\begin{array}{rrrrrrr} R_2 & 0 & 4 & -5 & 1 & 4 & 0 \\ -5 \cdot R_3 & 0 & 0 & 5 & -10 & 0 & -5 \\ \hline & 0 & 4 & 0 & -9 & 4 & -5 \end{array}$$

$\left[\begin{array}{ccc|ccc} 2 & 0 & 0 & 1 & 0 & 1 \\ 0 & 4 & 0 & -9 & 4 & -5 \\ 0 & 0 & -1 & 2 & 0 & 1 \end{array}\right]$

$$\begin{array}{rrrrrrr} -1 \cdot R_1 & 2 & 0 & 1 & -1 & 0 & 0 \\ R_3 & 0 & 0 & -1 & 2 & 0 & 1 \\ \hline & 2 & 0 & 0 & 1 & 0 & 1 \end{array}$$

$$\begin{bmatrix} 1 & 0 & 0 & \vdots & \dfrac{1}{2} & 0 & \dfrac{1}{2} \\[2mm] 0 & 1 & 0 & \vdots & -\dfrac{9}{4} & 1 & -\dfrac{5}{4} \\[2mm] 0 & 0 & 1 & \vdots & -2 & 0 & -1 \end{bmatrix}$$

Multiply $\dfrac{1}{2} \cdot R_1, \dfrac{1}{4} \cdot R_2$ and $-1 \cdot R_3$.

Thus, the inverse is $\begin{bmatrix} \dfrac{1}{2} & 0 & \dfrac{1}{2} \\[2mm] -\dfrac{9}{4} & 1 & -\dfrac{5}{4} \\[2mm] -2 & 0 & -1 \end{bmatrix}$.

(Inverse of a square matrix)

25. $2x + 5y = -17$

   $-x + y = -9$

The inverse of $\begin{bmatrix} 2 & 5 \\ -1 & 1 \end{bmatrix}$ is $\begin{bmatrix} \dfrac{1}{7} & -\dfrac{5}{7} \\[2mm] \dfrac{1}{7} & \dfrac{2}{7} \end{bmatrix}$

$$\begin{bmatrix} x \\ y \end{bmatrix} = \begin{bmatrix} \dfrac{1}{7} & -\dfrac{5}{7} \\[2mm] \dfrac{1}{7} & \dfrac{2}{7} \end{bmatrix} \begin{bmatrix} -17 \\ -9 \end{bmatrix}$$

Multiply the inverse times the column of numbers.

$$\begin{bmatrix} x \\ y \end{bmatrix} = \begin{bmatrix} -\dfrac{17}{7} + \dfrac{45}{7} \\[2mm] -\dfrac{17}{7} - \dfrac{18}{7} \end{bmatrix} = \begin{bmatrix} 4 \\ -5 \end{bmatrix}$$

$(4, -5)$ is the solution.

(Inverse of a square matrix)

26. $-2x - z = -1$

$\frac{1}{2}x + y - z = -6$

$4x + z = -3$

The inverse of $\begin{bmatrix} -2 & 0 & -1 \\ \frac{1}{2} & 1 & -1 \\ 4 & 0 & 1 \end{bmatrix}$ is $\begin{bmatrix} \frac{1}{2} & 0 & \frac{1}{2} \\ -\frac{9}{4} & 1 & -\frac{5}{4} \\ -2 & 0 & -1 \end{bmatrix}$

$\begin{bmatrix} x \\ y \\ z \end{bmatrix} = \begin{bmatrix} \frac{1}{2} & 0 & \frac{1}{2} \\ -\frac{9}{4} & 1 & -\frac{5}{4} \\ -2 & 0 & -1 \end{bmatrix} \begin{bmatrix} -1 \\ -6 \\ -3 \end{bmatrix} = \begin{bmatrix} -\frac{1}{2} + 0 - \frac{3}{2} \\ \frac{9}{4} - 6 + \frac{15}{4} \\ 2 + 0 + 3 \end{bmatrix} = \begin{bmatrix} -2 \\ 0 \\ 5 \end{bmatrix}$

The solution is $(-2, 0, 5)$.

(Inverse of a square matrix)

27. $\begin{vmatrix} 2 & 5 \\ 1 & 6 \end{vmatrix} = 2(6) - 1(5) = 7$ Use $ad - bc$.

(Determinants and Cramer's Rule)

28. $\begin{vmatrix} 4 & 2 \\ 3 & 0 \end{vmatrix} = 4(0) - 2(3) = -6$ Use $ad - bc$.

(Determinants and Cramer's Rule)

29. $\begin{vmatrix} 3 & 2 & 2 \\ -1 & -1 & 1 \\ 0 & 1 & 2 \end{vmatrix}$

$-6 + 0 + (-2) = -8$ Find the sum of the products going down the diagonals.

$0 + 3 + (-4) = -1$ Find the sum of the products going up the diagonals.

$-8-(-1) = -7$

Downward products minus upward products gives the determinant of –7.

(Determinants and Cramer's Rule)

30.
$$\begin{vmatrix} -1 & 6 & 3 \\ 2 & -4 & 1 \\ -3 & -2 & 5 \end{vmatrix}$$

$$(20-18-12)-(36+2+60) = -108$$

The value of the determinant is –108.

If you are required to use the cofactor and minor technique, it would look as follows for expansion about row 1:

$$(-1)^{1+1}(-1)\begin{vmatrix} -4 & 1 \\ -2 & 5 \end{vmatrix} + (-1)^{1+2}(6)\begin{vmatrix} 2 & 1 \\ -3 & 5 \end{vmatrix} + (-1)^{1+3}(3)\begin{vmatrix} 2 & -4 \\ -3 & -2 \end{vmatrix}$$

$$= 1(-1)(-20-(-2)) + (-1)(6)(10-(-3)) + 1(3)(-4-(12))$$

$$= -1(-18) - 6(13) + 3(-16) = -108$$

(Determinants and Cramer's Rule)

31. $2x + 3y = 5$

$-3x - 5y = -9$

$$x = \frac{\begin{vmatrix} 5 & 3 \\ -9 & -5 \end{vmatrix}}{\begin{vmatrix} 2 & 3 \\ -3 & -5 \end{vmatrix}} = \frac{-25-(-27)}{-10-(-9)} = \frac{2}{-1} = -2$$

$$y = \frac{\begin{vmatrix} 2 & 5 \\ -3 & -9 \end{vmatrix}}{-1} = \frac{-18-(-15)}{-1} = \frac{-3}{-1} = 3$$

(Determinants and Cramer's Rule)

32. $$x = \dfrac{\begin{vmatrix} 4 & 1 & 3 \\ -4 & 3 & -1 \\ -13 & -2 & -5 \end{vmatrix}}{\begin{vmatrix} 2 & 1 & 3 \\ 2 & 3 & -1 \\ 3 & -2 & -5 \end{vmatrix}} = \dfrac{66}{-66} = -1$$

$$y = \dfrac{\begin{vmatrix} 2 & 4 & 3 \\ 2 & -4 & -1 \\ 3 & -13 & -5 \end{vmatrix}}{\begin{vmatrix} 2 & 1 & 3 \\ 2 & 3 & -1 \\ 3 & -2 & -5 \end{vmatrix}}$$

$$z = \dfrac{\begin{vmatrix} 2 & 1 & 4 \\ 2 & 3 & -4 \\ 3 & -2 & -13 \end{vmatrix}}{\begin{vmatrix} 2 & 1 & 3 \\ 2 & 3 & -1 \\ 3 & -2 & -5 \end{vmatrix}}$$

(Determinants and Cramer's Rule)

33. $x + y \leq 2$
   $-x + y \leq 2$
      $x \geq 0$

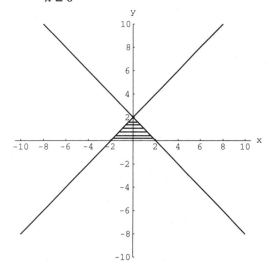

(Systems of inequalities)

34.　　$x^2 + y^2 \le 16$

　　　$x^2 + y^2 \ge 4$

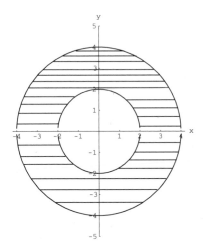

(Systems of inequalities)

35.　　$y \ge x^2$

　　　$y < \dfrac{1}{2}x + 2$

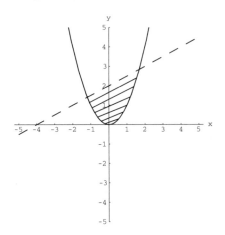

(Systems of inequalities)

36.　　Objective function: $C = 6x + 9y$

　　　Constraints:　　　$x \ge 0$

　　　　　　　　　　　$y \ge 0$

　　　　　　　　$4x + 3y \le 24$

　　　　　　　　$x + 2y \ge 6$

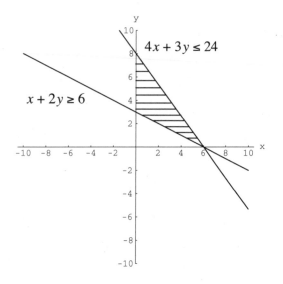

Vertices are $(0, 8)$, $(0, 3)$ and the intersection of the lines $4x + 3y = 24$ and $x + 2y = 6$:

$4x + 3y = 24$

$-4x - 8y = -24$ Multiply: $-4(x + 2y = 6)$

$-5y = 0$ Add the equations.

$y = 0$

$x = 6$

Check each vertex in the objective function:

$C = 6(0) + 9(8) = 72$

$C = 6(0) + 9(3) = 27$

$C = 6(6) + 9(0) = 36$

Therefore the maximum value of 72 occurs at $(0, 8)$ and the minimum value of 27 occurs at $(0, 3)$.

(Systems of inequalities)

37.   Use a table to organize the data:

	Model I	Model II
Cost	$100	$300
Profit	$100	$200
Number of units	x	y

$$x \geq 0$$ The merchant plans to stock the items.

$$y \geq 0$$

$$x + y \leq 100$$ Demand will not exceed 100 units.

$$100x + 300y \leq 20000$$ The merchant invests a maximum of $20,000 in inventory.

Maximum profit: $100x + 200y$ The merchant wishes to maximize profit.

Draw the graph:

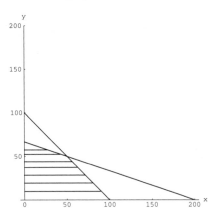

The vertices are (0, 66), (100, 0), and (50, 50), where (50, 50) is found by solving $x + y = 100$ and $100x + 300y = 20000$ simultaneously.

Check each vertex point in the profit expression:

(0, 66): $100(0) + 200(66) = 13{,}200$

(50, 50): $100(50) + 200(50) = 15{,}000$

(100, 0): $100(100) + 200(0) = 10{,}000$

Therefore, the maximum profit occurs when 50 hard drives of each type are ordered.

(Systems of inequalities)

Grade Yourself

Circle the question numbers that you had incorrect. Then indicate the number of questions you missed. If you answered more than three questions incorrectly, you will have to focus on that topic. If a topic has fewer than three questions and you had at least one wrong, we suggest you study that topic. Read your textbook or a review book or ask your teacher for help.

Subject: Systems of Equations, Matrices, and Systems of Inequalities

Topic	Question Numbers	Number Incorrect
Systems of equations	1, 2, 3, 4, 5, 6	
Linear systems	7, 8, 9, 10, 11, 12	
Augmented matrix	13, 14, 15, 16	
Matrix algebra	17, 18, 19, 20, 21	
Inverse of a square matrix	22, 23, 24, 25, 26	
Determinants and Cramer's Rule	27, 28, 29, 30, 31, 32	
Systems of inequalities	33, 34, 35, 36, 37	

Sequences and Other Topics

 ## Brief Yourself

This chapter contains questions about sequences, series, mathematical induction, the Binomial Theorem, and counting principles.

A **sequence** is an ordered list of numbers. Each number in the list is called a **term** of the sequence. We use subscripts to identify specific terms in a sequence. A **recursion formula** is a formula for finding the n-th term of a sequence using one or more earlier terms of the sequence. The sequence 3, 6, 9, 12 . . . has terms:

$$a_1 = 3, a_2 = 6, a_3 = 9, a_4 = 12, \ldots$$

This sequence could be defined recursively as:

$$a_1 = 3$$

$$a_n = a_{n-1} + 3 \quad \text{for } n > 1$$

To find a_2, n would equal 2 so

$$a_2 = a_{2-1} + 3 \qquad\qquad \text{Substitute } n = 2.$$

$$= a_1 + 3 \qquad\qquad \text{Simplify.}$$

$$= 3 + 3 = 6 \qquad\qquad \text{Replace } a_1 \text{ with 3.}$$

Note that each new term is found by adding the previous term and 3.

A **series** is a sum of the terms of a sequence. Special summation notation is used to write series.

$\displaystyle\sum_{i=1}^{4} (3i + 1)$ means replace i with integers from 1 to 4 in the formula $3i + 1$ and sum the terms. The **index of the summation**, i, is used to indicate the beginning and ending integer. In the formulas for the terms, i may be used as a term, factor, or exponent. Do not confuse the index i with the imaginary number i. The series we deal with do not involve imaginary numbers. The formula in a series may contain variables other than the index. These variables do *not* change as the index changes.

Summation notation can be used to write a sum of terms if we can find a formula for the terms. This usually requires a trial and error process; note that the answers are not unique.

An **arithmetic sequence** or **arithmetic progression** consists of terms that differ by a **common difference**, d. If the terms of an arithmetic sequence are listed, the common difference d can be found by subtracting any term from the term that follows it. Since the terms of an arithmetic sequence differ by d, the common difference, each term can be found by adding d to the previous term. The general term of an arithmetic sequence with first term a_1 and common difference d is

$$a_n = a_1 + (n - 1)d$$

If we know two terms of an arithmetic sequence, we can find a_1 and d by solving a system of two equations in two unknowns.

Sum of n Terms of an Arithmetic Sequence

The sum S_n of the first n terms of any arithmetic sequence with first term a_1, n-th term a_n, and common difference d is

$$S_n = \frac{n(a_1 + a_n)}{2} \quad \text{or } S_n = \frac{n}{2}[2a_1 + (n - 1)d]$$

We use the first formula when both the first term and n-th term are given or can be easily found.

A **geometric sequence** is a sequence where each term after the first term is a constant multiple of the preceding term. The constant multiple is called the **common ratio, r.** If the terms of a geometric sequence are listed, the common ratio r can be found by dividing any term by the term preceding it. We find terms of a geometric sequence by multiplying by r: $a_n = a_{n-1}r$, $n > 1$. The general term of a geometric sequence with first term a_1 and common ratio r is

$$a_n = a_1 r^{n-1}$$

We can substitute directly into the formula if we know a_1 and r.

A **geometric series** is the sum of the terms of a geometric sequence.

Sum of n Terms of a Geometric Sequence

The sum S_n of the first n terms of a geometric sequence with first term a_1 and common ratio r is

$$S_n = \frac{a_1(r^n - 1)}{r - 1} \qquad (r \neq 1)$$

The sum of the terms of certain infinite geometric series can be found. If the absolute value of the common ratio, $|r|$, is less than 1, the sum S of the infinite geometric series is $S = \dfrac{a_1}{1 - r}$.

To write a proof using mathematical induction, first show that the statement is true for $n = 1$. Then, assuming the statement is true for $n = k$, show that it is true for $n = k + 1$. If time allows on the test, begin by convincing yourself that the formula is true for several values of n. Several useful summation formulas (that can be proven by mathematical induction) are:

$$\sum_{n=1}^{n} n = \frac{n(n+1)}{2}$$

$$\sum_{n=1}^{n} n^2 = \frac{n(n+1)(2n+1)}{6}$$

$$\sum_{n=1}^{n} n^3 = \frac{n^2(n+1)^2}{4}$$

Definitions

$n! = n(n-1)(n-2) \ldots (2)(1)$ for n any positive integer

$0! = 1$

The **binomial coefficient** $\binom{n}{r}$ is defined by $\binom{n}{r} = \dfrac{n!}{r!(n-r)!}$

The Binomial Theorem

For x and y real numbers and n a positive integer,

$$(x+y)^n = \binom{n}{0} x^n y^0 + \binom{n}{1} x^{n-1} y^1 + \binom{n}{2} x^{n-2} y^2 + \ldots + \binom{n}{n} x^0 y^n$$

Take note of the following to help you remember the theorem:

1. The exponents on x start with n and decrease by 1 for each term.

2. The exponents on y start with 0 and increase by 1 for each term.

3. The binomial coefficient for each term is $\binom{n}{r}$ where r is the exponent on y.

4. The exponent on x plus the exponent on y equals n in each term.

Pascal's Triangle

The coefficients of the terms in the expansion of $(x + y)^n$ form an interesting pattern. When these coefficients are placed in rows, the display is called Pascal's Triangle:

			1			Coefficients in $(x+y)^0$
		1		1		Coefficients in $(x+y)^1$
	1		2		1	Coefficients in $(x+y)^2$
1		3		3		1 Coefficients in $(x+y)^3$
1	4	6	4	1		Coefficients in $(x+y)^4$

Each row begins and ends with 1 and each term inside is the sum of the two terms above it.

Finding the kth Term in a Binomial Expansion

The kth term in the expansion of $(x+y)^n$ is $\binom{n}{k-1} x^{n-(k-1)} y^{k-1}$

The Fundamental Counting Principle provides a formula for finding the number of ways events can occur. If E_1 can occur m_1 different ways and E_2 can occur m_2 different ways, then the two events can occur $m_1 \cdot m_2$ ways.

To calculate the number of ways r items can be selected from a total of n items, you will need the formulas for permuations and combination. When the order is important, use a *permutation*. That is, if n elements are taken r at a time, there are $_nP_r = \dfrac{n!}{(n-r)!}$ ways that this can occur. If the order in which the elements appear is *not* important, use $_nC_r = \dfrac{n!}{(n-r)!r!}$ to find the number of *combinations* of n items taken r at a time.

Some formulas you will need for working with probability are:

$P(E) = \dfrac{n(E)}{n(S)}$ where $P(E)$ is the probability of event E, $n(E)$ is the number of favorable outcomes of the event and $n(S)$ is the number of outcomes in the sample space.

$0 \le P(E) \le 1$

$P(A \cup B) = P(A) + P(B) - P(A \cap B)$

If A and B are mutually exclusive events, $P(A \cap B) = 0$

If A and B are independent events, $P(A \text{ and } B) = P(A) \cdot P(B)$

$P(A') = 1 - P(A)$ where A' is the complement of A.

Test Yourself

Find the first five terms of the sequence.

1. $a_n = 2n - 1$

2. $a_n = \dfrac{n+1}{n}$

3. $a_n = n^2$

4. $a_n = (-1)^{n+1}$

Write the first five terms of the sequence defined recursively.

5. $a_1 = 5 \quad a_n = -2a_{n-1} \quad n > 1$

6. $a_1 = 1 \quad a_n = 3a_{n-1} + 1 \quad n > 1$

7. $a_1 = 2 \quad a_n = 2a_{n-1} + n \quad n > 1$

Find the sum.

8. $\displaystyle\sum_{i=1}^{5} (2i - 3)$

9. $\displaystyle\sum_{i=2}^{6} i^2$

10. $\displaystyle\sum_{i=1}^{4} (i-1)(i+4)$

11. $\displaystyle\sum_{i=3}^{5} \dfrac{i^2 - 3}{2i + 4}$

Write out the terms in each series.

12. $\displaystyle\sum_{i=1}^{4} (2x - i)$

13. $\displaystyle\sum_{i=2}^{4} x^i$

Write in summation notation.

14. $9 + 16 + 25 + 36 + 49$

15. $1 + 8 + 27 + 64$

16. $\dfrac{1}{2} + \dfrac{2}{3} + \dfrac{3}{4} + \dfrac{4}{5} + \dfrac{5}{6}$

Find the common difference, d, for these arithmetic sequences.

17. $6, 11, 16, 21, 2 \ldots$

18. $4, 1, -2, -5, -8 \ldots$

19. $1, \dfrac{4}{3}, \dfrac{5}{3}, 2, \dfrac{7}{3}, \dfrac{8}{3}, 3 \ldots$

Write the first five terms for each arithmetic sequence with the given first term a_1 and common difference d.

20. $a_1 = 5 \quad d = 4$

21. $a_1 = 7 \quad d = -2$

Find the indicated term of the arithmetic sequence.

22. $a_1 = 6 \quad d = -4 \quad$ Find a_{23}.

23. $4, -2, -8, -14 \ldots \quad$ Find a_{100}.

24. Find a_{24} for the arithmetic sequence where $a_{11} = 35$ and $a_{15} = 47$.

25. Find the sum of the first 10 positive integers.

26. Find the sum of the first nine terms of the arithmetic sequence $3, 11, 19, 27, \ldots$

Find the common ratio, r, for these geometric sequences.

27. $3, 6, 12, 24, 48 \ldots$

28. $2, -6, 18, -54, 162 \ldots$

29. $4, 1, \dfrac{1}{4}, \dfrac{1}{16}, \dfrac{1}{64} \ldots$

Find the first five terms for each geometric sequence with the given first term a_1 and common ratio r.

30. $a_1 = 5, r = -2$

31. $a_1 = 1, r = \dfrac{1}{2}$

Find the indicated term of the geometric sequence.

32. $a_1 = 5 \quad r = -3 \quad$ Find a_4.

33. $1, 3, 9, 27, 81 \ldots \quad$ Find a_{10}.

Find a formula for the general term of each geometric sequence.

34. $3, 6, 12, 24 \ldots$

35. $\dfrac{1}{4}, \dfrac{3}{4}, \dfrac{9}{4}, \dfrac{27}{4} \ldots$

36. Find the general term of a geometric sequence with $a_2 = 18$ and $a_6 = 1{,}458$.

37. Find the sum of the first six terms of the geometric sequence with first term 8 and common ratio –2.

38. Find the sum of the first 10 terms of the geometric sequence $5, 10, 20, 40 \ldots$

39. Find the sum $\displaystyle\sum_{n=0}^{\infty} 5\left(\dfrac{1}{4}\right)^{n}$.

40. Find a fraction equivalent to the repeating decimal $0.454545 \ldots$

41. Use mathematical induction to prove the formula for every positive integer n: $1 + 3 + 5 + 7 + \ldots + (2n - 1) = n^2$

42. Use mathematical induction to prove the formula for every positive integer n: $1(2) + 2(3) + 3(4) + \ldots + n(n + 1) = \dfrac{n(n + 1)(n + 2)}{3}$

Find the sum using the formulas for the sums of powers of integers.

43. $\displaystyle\sum_{n=1}^{100} n$

44. $\displaystyle\sum_{n=1}^{5} n^2$

45. $\displaystyle\sum_{n=1}^{5} n^3$

Evaluate.

46. $\dfrac{4!}{0!}$

47. $\dfrac{8!}{3!5!}$

48. $\dfrac{6!}{6!0!}$

Evaluate.

49. $\dbinom{4}{0}$

50. $\dbinom{4}{1}$

51. $\dbinom{4}{2}$

52. $\dbinom{4}{3}$

53. $\dbinom{4}{4}$

54. Expand $(x + y)^4$.

55. Expand $(2a + 3)^5$.

56. Use Pascal's Triangle to expand $(x + y)^6$.

57. Find the fifth term in the expansion of $(x + y)^9$.

58. Find the eighth term in the expansion of $(3a - 2b)^{12}$.

59. To dress for the party, Adam can choose one of two suits, one of four shirts, and one of six ties. Determine the number of possible outfits, assuming none of the choices clash with other choices.

60. In how many different ways can a 10-question true-false exam be answered? (Assume no questions are omitted.)

61. How many automobile license plates can a state issue if each plate consists of two letters followed by four digits?

62. The password for a particular computer program consists of six entries, where the first three entries must be lower case letters, the next entry must be a digit from 1 to 9, and the last two entries may be upper or lower case letters. How many distinct passwords are possible?

63. Study groups of 4 students each are to be formed from a class of 24 students. How many different groups of 4 students can be formed?

64. There are 50 numbers in the state's lottery game. In how many ways can a player select 6 of the numbers?

65. Determine the sample space for the experiment: A coin is tossed and then a six-sided die is rolled.

66. Find the probability of drawing a red face card from a standard deck of cards.

67. Of the 46 students enrolled in a mathematics class, 30 students plan to take another mathematics course. Of these, 12 are planning to go to graduate school. Of the other 16 students, 3 are planning to go to graduate school. If a student is selected at random, what is the probability that the person is (a) planning to go to graduate school, (b) not planning to go to graduate school, (c) planning to take another mathematics course but not planning to go to graduate school?

 # Check Yourself

1. $a_n = 2n - 1$

   $a_1 = 2(1) - 1 = 2 - 1 = 1$ Replace n with 1, 2, 3, 4, and 5.

   $a_2 = 2(2) - 1 = 4 - 1 = 3$

   $a_3 = 2(3) - 1 = 6 - 1 = 5$

   $a_4 = 2(4) - 1 = 8 - 1 = 7$

   $a_5 = 2(5) - 1 = 10 - 1 = 9$

   (Sequences)

2. $a_n = \dfrac{n + 1}{n}$

   $a_1 = \dfrac{1 + 1}{1} = \dfrac{2}{1}$ Replace n with 1, 2, 3, 4, and 5.

   $a_2 = \dfrac{2 + 1}{2} = \dfrac{3}{2}$

   $a_3 = \dfrac{3 + 1}{3} = \dfrac{4}{3}$

   $a_4 = \dfrac{4 + 1}{4} = \dfrac{5}{4}$

   $a_5 = \dfrac{5 + 1}{5} = \dfrac{6}{5}$

   (Sequences)

3. $a_n = n^2$

   $a_1 = 1^2 = 1$ Replace n with 1, 2, 3, 4, and 5.

   $a_2 = 2^2 = 4$

$$a_3 = 3^2 = 9$$

$$a_4 = 4^2 = 16$$

$$a_5 = 5^2 = 25$$

(Sequences)

4. $a_n = (-1)^{n+1}$

$a_1 = (-1)^{1+1} = (-1)^2 = 1$ Replace n with 1, 2, 3, 4, and 5.

$a_2 = (-1)^{2+1} = (-1)^3 = -1$

$a_3 = (-1)^{3+1} = (-1)^4 = 1$

$a_4 = (-1)^{4+1} = (-1)^5 = -1$

$a_5 = (-1)^{5+1} = (-1)^6 = 1$

(Sequences)

5. $a_1 = 5$ a_1 is the first term.

$a_n = -2a_{n-1}$ This is the recursion formula.

$a_2 = -2a_{2-1}$ Substitute $n = 2$.

$\quad = -2a_1$ Simplify.

$\quad = -2(5)$ Replace a_1 with 5.

$\quad = -10$ a_2 is the second term.

$a_3 = -2\,a_{3-1}$ Substitute $n = 3$.

$\quad = -2a_2$ Simplify.

$\quad = -2(-10)$ Replace a_2 with -10.

$\quad = 20$ This is the third term.

$a_4 = -2a_{4-1}$ Substitute $n = 4$.

$\quad = -2a_3$ Simplify.

$\quad = -2(20)$ Replace a_3 with 20.

$\quad = -40$ This is the fourth term.

$a_5 = -2a_{5-1}$ Substitute $n = 5$.

$\quad = -2a_4$ Simplify.

$= -2(-40)$	Replace a_4 with -40.
$= 80$	This is the fifth term.

(Sequences)

6. $a_1 = 1$ — a_1 is the first term.

$a_n = 3a_{n-1} + 1$	This is the recursion formula.
$a_2 = 3a_{2-1} + 1$	Substitute $n = 2$.
$= 3a_1 + 1$	Simplify.
$= 3(1) + 1$	Replace a_1 with 1.
$= 4$	This is the second term.
$a_3 = 3a_{3-1} + 1$	Substitute $n = 3$.
$= 3a_2 + 1$	Simplify.
$= 3(4) + 1$	Replace a_2 with 4.
$= 13$	This is the third term.
$a_4 = 3a_{4-1} + 1$	Substitute $n = 4$.
$= 3a_3 + 1$	Simplify.
$= 3(13) + 1$	Replace a_3 with 13.
$= 40$	This is the fourth term.
$a_5 = 3a_{5-1} + 1$	Substitute $n = 5$.
$= 3a_4 + 1$	Simplify.
$= 3(40) + 1$	Replace a_4 with 40.
$= 121$	This is the fifth term.

(Sequences)

7. $a_1 = 2$

$a_n = 2a_{n-1} + n$	This is the recursion formula.
$a_2 = 2a_{2-1} + 2$	Substitute $n = 2$.
$= 2a_1 + 2$	Simplify.
$= 2(2) + 2$	Replace a_1 with 2.
$= 6$	This is the second term.
$a_3 = 2a_{3-1} + 3$	Substitute $n = 3$.
$= 2a_2 + 3$	Simplify.

$= 2(6) + 3$	Replace a_2 with 6.
$= 15$	This is the third term.
$a_4 = 2a_{4-1} + 4$	Substitute $n = 4$.
$= 2a_3 + 4$	Simplify.
$= 2(15) + 4$	Replace a_3 with 15.
$= 34$	This is the fourth term.
$a_5 = 2a_{5-1} + 5$	Substitute $n = 5$.
$= 2a_4 + 5$	Simplify.
$= 2(34) + 5$	Replace a_4 with 34.
$= 73$	This is the fifth term.

(Sequences)

8. $\displaystyle\sum_{i=1}^{5} (2i - 3)$

$= [2(1) - 3] + [2(2) - 3] + [2(3) - 3] + [2(4) - 3] + [2(5) - 3]$

	Replace i, starting with 1 and ending with 5.
$= -1 + 1 + 3 + 5 + 7$	Simplify each expression in brackets.
$= 15$	Add.

(Series)

9. $\displaystyle\sum_{i=2}^{6} i^2 = (2)^2 + (3)^2 + (4)^2 + (5)^2 + (6)^2$ — Replace i, starting with 2 and ending with 6.

$= 4 + 9 + 16 + 25 + 36$	Simplify each term.
$= 90$	Add.

(Series)

10. $\displaystyle\sum_{i=1}^{4} (i-1)(i+4) = (1-1)(1+4) + (2-1)(2+4) + (3-1)(3+4) + (4-1)(4+4)$

	Replace i, starting with 1 and ending with 4.
$= 0(5) + 1(6) + 2(7) + 3(8)$	Simplify.
$= 0 + 6 + 14 + 24$	
$= 44$	Add.

(Series)

11. $\displaystyle\sum_{i=3}^{5} \frac{i^2-3}{2i+4} = \frac{3^2-3}{2(3)+4} + \frac{4^2-3}{2(4)+4} + \frac{5^2-3}{2(5)+4}$ Replace i, starting with 3 and ending with 5.

$\displaystyle = \frac{9-3}{6+4} + \frac{16-3}{8+4} + \frac{25-3}{10+4}$ Perform multiplications.

$\displaystyle = \frac{6}{10} + \frac{13}{12} + \frac{22}{14}$ Simplify numerators and denominators.

$\displaystyle = \frac{3}{5} + \frac{13}{12} + \frac{11}{7}$ Reduce.

$\displaystyle = \frac{3}{5}\cdot\frac{84}{84} + \frac{13}{12}\cdot\frac{35}{35} + \frac{11}{7}\cdot\frac{60}{60}$ Use $5 \cdot 12 \cdot 7 = 420$ as the LCD.

$\displaystyle = \frac{252}{420} + \frac{455}{420} + \frac{660}{420} = \frac{1367}{420}$ Add.

(Series)

12. $\displaystyle\sum_{i=1}^{4}(2x-i) = (2x-1)+(2x-2)+(2x-3)+(2x-4)$ Replace i starting with 1 and ending with 4.

Stop here since the directions only ask for the terms to be written out.

(Series)

13. $\displaystyle\sum_{i=2}^{4} x^i = x^2 + x^3 + x^4$ Replace i, starting with 2 and ending with 4.

(Series)

14. $9 + 16 + 25 + 36 + 49 =$ These are perfect squares.

$\displaystyle 3^2 + 4^2 + 5^2 + 6^2 + 7^2 = \sum_{i=3}^{7} i^2$

(Series)

15. $1 + 8 + 27 + 64 =$ These are perfect cubes.

$\displaystyle 1^3 + 2^3 + 3^3 + 4^3 = \sum_{i=1}^{4} i^3$

(Series)

16. $\displaystyle\frac{1}{2} + \frac{2}{3} + \frac{3}{4} + \frac{4}{5} + \frac{5}{6} = \sum_{i=1}^{5} \frac{i}{i+1}$ The numerators are successive integers, starting at 1.

Each denominator is one more than the numerator.

(Series)

17. $6, 11, 16, 21, 26 \ldots$

   $26 - 21 = 5$ Subtract any term from the term that follows it.

   $d = 5$

(Arithmetic sequences and series)

18. $4, 1, -2, -5, -8 \ldots$

   $-5 - (-2) = -5 + 2 = -3$ Subtract any term from the term that follows it.

   $d = -3$

(Arithmetic sequences and series)

19. $1, \dfrac{4}{3}, \dfrac{5}{3}, 2, \dfrac{7}{3}, \dfrac{8}{3}, 3 \ldots$

   $\dfrac{5}{3} - \dfrac{4}{3} = \dfrac{1}{3}$ Subtract any term from the term that follows it.

   $d = \dfrac{1}{3}$

(Arithmetic sequences and series)

20. $a_1 = 5$ $d = 4.$

   $a_2 = 5 + 4 = 9$ Add 4 to a_1.

   $a_3 = 9 + 4 = 13$ Add 4 to a_2.

   $a_4 = 13 + 4 = 17$ Add 4 to a_3.

   $a_5 = 17 + 4 = 21$ Add 4 to a_4.

(Arithmetic sequences and series)

21. $a_1 = 7$ $d = -2.$

   $a_2 = 7 + (-2) = 5$ Add -2 to a_1.

   $a_3 = 5 + (-2) = 3$ Add -2 to a_2.

   $a_4 = 3 + (-2) = 1$ Add -2 to a_3.

   $a_5 = 1 + (-2) = -1$ Add -2 to a_4.

(Arithmetic sequences and series)

22. $a_1 = 6$ $d = -4.$

   $a_n = a_1 + (n - 1)d$ Write the formula.

   $a_{23} = a_1 + (23 - 1)d$ $n = 23$

   $a_{23} = a_1 + 22d$ Simplify.

$a_{23} = 6 + 22(-4)$ Substitute $a_1 = 6$ and $d = -4$.

$a_{23} = 6 - 88$ Multiply.

$a_{23} = -82$ Subtract.

(Arithmetic sequences and series)

23. In $4, -2, -8, -14\ldots$, $a_1 = 4$ and $d = -8 - (-2) = -8 + 2 = -6$

$a_n = a_1 + (n-1)d$ Write the formula.

$a_{100} = a_1 + (100-1)d$ $n = 100$

$a_{100} = a_1 + 99d$ Simplify.

$a_{100} = 4 + 99(-6)$ Substitute $a_1 = 4$ and $d = -6$.

$a_{100} = 4 - 594$ Multiply.

$a_{100} = -590$ Subtract.

(Arithmetic sequences and series)

24. $a_{11} = a_1 + (11-1)d$ Write the formula for a_n with $n = 11$ since a_{11} is given.

$a_{11} = a_1 + 10d$ Simplify.

$35 = a_1 + 10d$ Substitute $a_{11} = 35$.

$a_{15} = a_1 + (15-1)d$ Write the formula for a_n with $n = 15$.

$a_{15} = a_1 + 14d$ Simplify.

$47 = a_1 + 14d$ Substitute $a_{15} = 47$.

Now solve the resulting system of equations:

$a_1 + 10d = 35$ Solve the system of equations.
$a_1 + 14d = 47$

$-a_1 - 10d = -35$ Multiply the first equation by -1.
$a_1 + 14d = 47$

$4d = 12$ Add the equations.

$d = 3$ Divide by 4.

$a_1 + 10(3) = 35$ Substitute $d = 3$ into the first equation.

$a_1 + 30 = 35$

$a_1 = 5$

$$a_{24} = a_1 + (24 - 1)d$$ Use the formula for a_n with $n = 24$.

$$a_{24} = 5 + 23(3)$$ Substitute $a_1 = 5$ and $d = 3$.

$$a_{24} = 5 + 69 = 74$$

(Arithmetic sequences and series)

25. $a_1 = 1, a_{10} = 10$ Here $n = 10$ so find a_1 and a_{10}.

$$S_n = \frac{n(a_1 + a_n)}{2}$$ Write the formula for S_n.

$$S_{10} = \frac{10(1 + 10)}{2}$$ Substitute $n = 10$, $a_1 = 1$, and $a_{10} = 10$.

$$S_{10} = \frac{10(11)}{2} = 55$$ Simplify.

Or use the alternate formula:

$$S_n = \frac{n}{2}[2a_1 + (n - 1)d]$$ Write the formula for S_n.

$$S_{10} = \frac{10}{2}[2(1) + (10 - 1)1]$$ Substitute $n = 10$, $a_1 = 1$, $d = 1$.

$$S_{10} = \frac{10}{2}[2 + 9]$$

$$S_{10} = 5(11) = S_{10} = 55$$

(Arithmetic sequences and series)

26.   We know $n = 9$ but do not know a_9, the ninth term. We could find a_9, but it is more convenient to use the

formula $S_n = \frac{n}{2}[2a_1 + (n - 1)d]$.

$$S_9 = \frac{9}{2}[2(3) + (9 - 1)8]$$ Substitute $n = 9$, $a_1 = 3$, $d = 8$.

$$S_9 = \frac{9}{2}[2(3) + 8(8)]$$ Simplify.

$$S_9 = \frac{9}{2}[6 + 64]$$ Multiply.

$$S_9 = \frac{9}{2}(70)$$ Add.

$$S_9 = 315$$

(Arithmetic sequences and series)

27. $3, 6, 12, 24, 48, \ldots$

$\dfrac{24}{12} = 2$ Divide any term by the term preceding it.

$\dfrac{12}{6} = 2$

$r = 2$

(Geometric sequences and series)

28. $2, -6, 18, -54, 162, \ldots$

$\dfrac{-54}{18} = -3$ Divide any term by the term preceding it.

$\dfrac{18}{-6} = -3$

$r = -3$

(Geometric sequences and series)

29. $4, 1, \dfrac{1}{4}, \dfrac{1}{16}, \dfrac{1}{64}, \ldots$

$1 \div 4 = \dfrac{1}{4}$ Divide any term by the term preceding it.

$\dfrac{1}{16} \div \dfrac{1}{4} = \dfrac{1}{16} \cdot \dfrac{4}{1} = \dfrac{1}{4}$

$r = \dfrac{1}{4}$

(Geometric sequences and series)

30. $a_1 = 5, r = -2$

$a_2 = a_1 r = 5(-2) = -10$ Multiply a_1 by $r = -2$.

$a_3 = a_2 r = -10(-2) = 20$ Multiply a_2 by $r = -2$.

$a_4 = a_3 r = 20(-2) = -40$ Multiply a_3 by $r = -2$.

$a_5 = a_4 r = -40(-2) = 80$ Multiply a_4 by $r = -2$.

(Geometric sequences and series)

31. $a_1 = 1, r = \dfrac{1}{2}$

$a_2 = a_1 r = 1\left(\dfrac{1}{2}\right) = \dfrac{1}{2}$ Multiply a_1 by $r = \dfrac{1}{2}$.

$$a_3 = a_2 r = \frac{1}{2}\left(\frac{1}{2}\right) = \frac{1}{4}$$ Multiply a_2 by $r = \frac{1}{2}$.

$$a_4 = \frac{1}{4}\left(\frac{1}{2}\right) = \frac{1}{8}$$ Multiply a_3 by $r = \frac{1}{2}$.

$$a_5 = \frac{1}{8}\left(\frac{1}{2}\right) = \frac{1}{16}$$ Multiply a_4 by $r = \frac{1}{2}$.

(Geometric sequences and series)

32. $a_n = a_1 r^{n-1}$ Write the formula.

       $a_4 = a_1 r^{4-1}$ Substitute $n = 4$.

       $a_4 = a_1 r^3$ Simplify.

       $a_4 = 5(-3)^3$ Substitute $a_1 = 5$ and $r = -3$.

       $a_4 = 5(-27) = -135$

(Geometric sequences and series)

33.   In the geometric sequence $1, 3, 9, 27, 81 \ldots$, $a_1 = 1$ and $r = \frac{3}{1} = 3$ (divide any term by the preceding term).

       $a_n = a_1 r^{n-1}$ Write the formula.

       $a_{10} = a_1 r^{10-1}$ Substitute $n = 10$.

       $a_{10} = a_1 r^9$ Simplify.

       $a_{10} = 1(3)^9$ Substitute $a_1 = 1$ and $r = 3$.

       $a_{10} = 3^9 = 19{,}683$

(Geometric sequences and series)

34.   In $3, 6, 12, 24 \ldots$, $a_1 = 3$, $r = \frac{6}{3} = 2$

       $a_n = a_1 r^{n-1}$ Write the formula.

       $a_n = 3(2)^{n-1}$ Substitute $a_1 = 3$, $r = 2$.

(Geometric sequences and series)

35.   In $\frac{1}{4}, \frac{3}{4}, \frac{9}{4}, \frac{27}{4} \ldots$, $a_1 = \frac{1}{4}$, $r = \frac{3}{4} \div \frac{1}{4} = \frac{3}{4} \cdot \frac{4}{1} = 3$

       $a_n = a_1 r^{n-1}$ Write the formula.

$$a_n = \frac{1}{4}(3)^{n-1}$$

(Geometric sequences and series)

36. $a_n = a_1 r^{n-1}$ Write the formula.

$a_2 = a_1 r^{2-1}$ Substitute $n = 2$ since a_2 is given.

$a_2 = a_1 r^1$ Simplify.

$18 = a_1 r^1$ Substitute $a_2 = 18$.

$a_n = a_1 r^{n-1}$ Write the formula.

$a_6 = a_1 r^{6-1}$ Substitute $n = 6$ since a_6 is given.

$a_6 = a_1 r^5$ Simplify.

$1{,}458 = a_1 r^5$ Substitute $a_6 = 1{,}458$.

Now set up a ratio using the two equalities.

$$\frac{a_1 r^5}{a_1 r^1} = \frac{1458}{18}$$

$r^4 = 81$ Simplify each side.

$r = 3$ Take the fourth root of both sides.

Now find a_1:

$18 = a_1 r^1$

$18 = a_1 (3)^1$

$6 = a_1$

So the general term a_n is

$a_n = 6(3)^{n-1}$.

(Geometric sequences and series)

37. $S_6 = \dfrac{a_1(r^6 - 1)}{r - 1}$ Use $n = 6$.

$S_6 = \dfrac{8[(-2)^6 - 1]}{(-2) - 1}$ Substitute $a_1 = 8$ and $r = -2$.

$$S_6 = \frac{8(64-1)}{-3}$$ 　　　　　　Simplify.

$$S_6 = -168$$ 　　　　　　Reduce.

(Geometric sequences and series)

38.　In the sequence 5, 10, 20, 40 . . . , $a_1 = 5, r = \dfrac{10}{5} = 2$.

$$S_{10} = \frac{a_1(r^{10}-1)}{r-1}$$ 　　　　　　Use $n = 10$.

$$S_{10} = \frac{5(2^{10}-1)}{2-1}$$ 　　　　　　Substitute $a_1 = 5, r = 2$.

$$S_{10} = 5(2^{10}-1) = 5{,}115$$ 　　　　　　Simplify.

(Geometric sequences and series)

39.　$S = \dfrac{a_1}{1-r}$ 　　　　　　Write the formula.

$$S = \frac{5}{1-\frac{1}{4}}$$ 　　　　　　Substitute $a_1 = 5, r = \dfrac{1}{4}$.

$$= \frac{5}{\frac{3}{4}}$$ 　　　　　　Simplify.

$$= 5 \cdot \frac{4}{3} = \frac{20}{3}$$ 　　　　　　Invert and multiply.

(Geometric sequences and series)

40.　Write the decimal as a sum: $0.454545\ldots = 0.45 + 0.0045 + 0.000045 + \ldots$

The terms in the sum are a geometric sequence with $a_1 = 0.45$ and $r = \dfrac{0.0045}{0.45} = 0.01$.

$$S = \frac{a_1}{1-r}$$ 　　　　　　Write the formula.

$$S = \frac{0.45}{1-0.01}$$ 　　　　　　Substitute $a_1 = 0.45$ and $r = 0.01$.

$$S = \frac{0.45}{0.99}$$ 　　　　　　Simplify.

$$S = \frac{45}{99} = \frac{5}{11}$$ 　　　　　　Reduce. Note that the division of 5 by 11 produces the repeated decimal $0.454545\ldots$

(Geometric sequences and series)

41. To prove $1 + 3 + 5 + 7 + \ldots + (2n - 1) = n^2$, begin by showing that the statement is true for $n = 1$:

$1 = (1)^2$ is true.

Assume the statement is true for $n = k$:

$1 + 3 + 5 + 7 + \ldots + (2k - 1) = k^2$

Show that statement is true for $n = k + 1$:

$1 + 3 + 5 + 7 + \ldots + (2(k + 1) - 1) = (k + 1)^2$

Note that $2((k + 1) - 1) = 2k + 1$, which is the $k + 1$ term, will be added to each side of the statement we assumed to be true.

Begin with the statement *assumed* to be true:

$1 + 3 + 5 + 7 + \ldots + (2k - 1) = k^2$

$1 + 3 + 5 + 7 + \ldots + (2k - 1) + (2k + 1) = k^2 + (2k + 1)$ Add $2k + 1$ to each side of the equation.

$1 + 3 + 5 + 7 + \ldots + (2k - 1) + (2k + 1) = (k + 1)^2$ Factor the right side of the equation.

Notice that the statement we wished to show true has been obtained, and the proof is complete.

(Mathematical induction)

42. To prove $1(2) + 2(3) + 3(4) + \ldots + n(n + 1) = \dfrac{n(n + 1)(n + 2)}{3}$ is true, begin by showing that the statement is true for $n = 1$:

$1(2) = \dfrac{1(1 + 1)(1 + 2)}{3}$ which is true.

Assume:

$1(2) + 2(3) + 3(4) + \ldots + k(k + 1) = \dfrac{k(k + 1)(k + 2)}{3}$

Show:

$1(2) + 2(3) + 3(4) + \ldots + (k + 1)(k + 2) = \dfrac{(k + 1)(k + 2)(k + 3)}{3}$

Begin with the statement *assumed* to be true:

$1(2) + 2(3) + 3(4) + \ldots + k(k + 1) = \dfrac{k(k + 1)(k + 2)}{3}$

Add $(k + 1)(k + 2)$ to each side of the equation:

$1(2) + 2(3) + 3(4) + \ldots + k(k + 1) + (k + 1)(k + 2) = \dfrac{k(k + 1)(k + 2)}{3} + (k + 1)(k + 2)$

$= \dfrac{k(k + 1)(k + 2)}{3} + \dfrac{3(k + 1)(k + 2)}{3}$

$$= \frac{k(k+1)(k+2) + 3(k+1)(k+2)}{3}$$

$$= \frac{(k+1)(k+1)[k+3]}{3}$$

$$= \frac{(k+1)(k+2)(k+3)}{3}$$

Notice that the statement we wished to show true has been obtained, and the proof is complete.

(Mathematical induction)

43. $\displaystyle\sum_{n=1}^{100} n = \frac{100(100+1)}{2}$ Use $\displaystyle\sum_{n=1}^{n} n = \frac{n(n+1)}{2}$

$$= 50(101) = 5050$$

(Mathematical induction)

44. $\displaystyle\sum_{n=1}^{5} n^2 = \frac{5(5+1)(2(5)+1)}{6}$ Use $\displaystyle\sum_{n=1}^{n} n^2 = \frac{n(n+1)(2n+1)}{6}$

$$= \frac{5(6)(11)}{6} = 55$$

(Mathematical induction)

45. $\displaystyle\sum_{n=1}^{5} n^3 = \frac{5^2(5+1)^2}{4}$ Use $\displaystyle\sum_{n=1}^{n} n^3 = \frac{n^2(n+1)^2}{4}$

$$= \frac{25(36)}{4} = 225$$

(Mathematical induction)

46. $\displaystyle\frac{4!}{0!} = \frac{4 \cdot 3 \cdot 2 \cdot 1}{1} = 24$ Use the definition of $n!$ and $0!$

(The Binomial Theorem)

47. $\displaystyle\frac{8!}{3!5!} = \frac{8 \cdot 7 \cdot 6 \cdot 5 \cdot 4 \cdot 3 \cdot 2 \cdot 1}{3 \cdot 2 \cdot 1 \cdot 5 \cdot 4 \cdot 3 \cdot 2 \cdot 1} = 56$ Use the definition of $n!$

(The Binomial Theorem)

48. $\displaystyle\frac{6!}{6!0!} = \frac{6 \cdot 5 \cdot 4 \cdot 3 \cdot 2 \cdot 1}{6 \cdot 5 \cdot 4 \cdot 3 \cdot 2 \cdot 1 \cdot 1} = 1$ Use the definition of $n!$ and $0!$

(The Binomial Theorem)

49. $\dbinom{4}{0} = \dfrac{4!}{0!(4-0)!}$ Use the definition of $\dbinom{n}{r}$ with $n = 4$, $r = 0$.

$= \dfrac{4!}{0!4!}$ Simplify.

$= \dfrac{4!}{1 \cdot 4!} = 1$ Use $\dfrac{n!}{n!} = 1$.

(The Binomial Theorem)

50. $\dbinom{4}{1} = \dfrac{4!}{1!(4-1)!}$ Use the definition of $\dbinom{n}{r}$ with $n = 4$, $r = 1$.

$= \dfrac{4!}{1!3!}$ Simplify.

$= \dfrac{4 \cdot 3 \cdot 2 \cdot 1}{1 \cdot 3 \cdot 2 \cdot 1} = 4$ Use the definition of $n!$

(The Binomial Theorem)

51. $\dbinom{4}{2} = \dfrac{4!}{2!(4-2)!}$ Use the definition of $\dbinom{n}{r}$ with $n = 4$, $r = 2$.

$= \dfrac{4!}{2!2!}$ Simplify.

$= \dfrac{4 \cdot 3 \cdot 2 \cdot 1}{2 \cdot 1 \cdot 2 \cdot 1} = 6$ Use the definition of $n!$

(The Binomial Theorem)

52. $\dbinom{4}{3} = \dfrac{4!}{3!(4-3)!}$ Use the definition of $\dbinom{n}{r}$ with $n = 4$, $r = 3$.

$= \dfrac{4!}{3!1!} = \dfrac{4 \cdot 3 \cdot 2 \cdot 1}{3 \cdot 2 \cdot 1 \cdot 1} = 4$ Use the definition of $n!$

(The Binomial Theorem)

53. $\dbinom{4}{4} = \dfrac{4!}{4!(4-4)!}$ Use the definition of $\dbinom{n}{r}$ with $n = 4$, $r = 4$.

$= \dfrac{4!}{4!0!} = \dfrac{4!}{4!1} = 1$ Use $0! = 1$.

(The Binomial Theorem)

54.   Use the formula with $n = 4$:

$$\dbinom{4}{0} x^4 y^0 + \dbinom{4}{1} x^3 y^1 + \dbinom{4}{2} x^2 y^2 + \dbinom{4}{3} x^1 y^3 + \dbinom{4}{4} x^0 y^4$$

Note that the coefficients were computed in problems 49 through 53.

$$= 1x^4 + 4x^3y + 6x^2y^2 + 4xy^3 + 1y^4$$

(The Binomial Theorem)

55. Use the formula with $n = 5$ substituting $2a$ for x and 3 for y:

$$\binom{5}{0}(2a)^5(3)^0 + \binom{5}{1}(2a)^4(3)^1 + \binom{5}{2}(2a)^3(3)^2 + \binom{5}{3}(2a)^2(3)^3 + \binom{5}{4}(2a)^1(3)^4 + \binom{5}{5}(2a)^0(3)^5$$

$$= \frac{5!}{0!(5-0)!}32a^5 + \frac{5!}{1!(5-1)!}(16a^4)(3) + \frac{5!}{2!(5-2)!}(8a^3)(9) + \frac{5!}{3!(5-3)!}(4a^2)(27) +$$

$$\frac{5!}{4!(5-4)!}(2a)(81) + \frac{5!}{5!(5-5)!}(1)(243)$$

$$= (1)32a^5 + 5(16a^4)(3) + 10(8a^3)(9) + 10(4a^2)(27) + 5(2a)(81) + 1(1)(243)$$

$$= 32a^5 + 240a^4 + 720a^3 + 1080a^2 + 810a + 243$$

(The Binomial Theorem)

56. Building on the fourth row, we have

$$\begin{array}{ccccccc} 1 & 4 & 6 & 4 & 1 & & \\ 1 & 5 & 10 & 10 & 5 & 1 & \\ 1 & 6 & 15 & 20 & 15 & 6 & 1 \end{array}$$

so,

$$(x + y)^6 = 1x^6 + 6x^5y + 15x^4y^2 + 20x^3y^3 + 15x^2y^4 + 6xy^5 + 1y^6$$

(The Binomial Theorem)

57. $\binom{n}{k-1}x^{n-(k-1)}y^{k-1}$ Write the formula.

$\binom{9}{5-1}x^{9-(5-1)}y^{5-1}$ Substitute $n = 9, k = 5$.

$\binom{9}{4}x^5y^4$ Simplify.

$\dfrac{9!}{4!(9-4)!}x^5y^4$ Find the coefficient.

$\dfrac{9 \cdot 8 \cdot 7 \cdot 6 \cdot 5!}{4 \cdot 3 \cdot 2 \cdot 1 \cdot 5!}x^5y^4$ Simplify.

$= 126x^5y^4$

(The Binomial Theorem)

58. $\binom{n}{k-1}x^{n-(k-1)}y^{k-1}$ Write the formula.

$\binom{12}{8-1}(3a)^{12-(8-1)}(-2b)^{8-1}$ Substitute $n = 12, k = 8, x = 3a$ and $y = -2b$.

$$\binom{12}{7} (3a)^5(-2b)^7 \qquad\qquad \text{Simplify.}$$

$$\frac{12!}{7!(12-7)!} (243a^5)(-128b^7) = \frac{12 \cdot 11 \cdot 10 \cdot 9 \cdot 8 \cdot 7!}{7!5!} (-31104a^5b^7) = 792(-31104a^5b^7)$$

$$= -24{,}634{,}368a^5b^7$$

(The Binomial Theorem)

59.　Use the Fundamental Counting Principle: $2 \cdot 4 \cdot 6 = 48$ possible outfits.

(Counting principles)

60.　Use the Fundamental Counting Principle: $2 \cdot 2 \cdot 2 \cdot 2 \cdot 2 \cdot 2 \cdot 2 \cdot 2 \cdot 2 \cdot 2 = 2^{10} = 1{,}024.$

(Counting principles)

61.　Number of license plates $= 26 \cdot 26 \cdot 10 \cdot 10 \cdot 10 \cdot 10 = 6{,}760{,}000.$

(Counting principles)

62.　$26 \cdot 26 \cdot 26 \cdot 9 \cdot 52 \cdot 52 = 427{,}729{,}536$ where the last two entries are 52 to account for 26 lower case letter possibilities and 26 upper case letter possibilities.

(Counting principles)

63.　You are asked to find the number of combinations of 24 items taken 4 at a time.

$$_{24}C_4 = \frac{24!}{(24-4)!4!} = \frac{24!}{20!4!} = 10{,}626.$$

(Counting principles)

64.　$_{50}C_6 = \dfrac{50!}{(50-6)!6!} = 15{,}890{,}700$

(Counting principles)

65.　Let H stand for heads and T for tails. Then the sample space is:

$\{(H, 1), (H, 2), (H, 3), (H, 4), (H, 5), (H, 6), (T, 1), (T, 2), (T, 3), (T, 4), (T, 5), (T, 6)\}.$

(Probability)

66.　$P(\text{red face card}) = \dfrac{\text{number of red face cards}}{\text{number of cards in the deck}} = \dfrac{6}{52} = \dfrac{3}{26}$

(Probability)

67. (a) P(planning to go to graduate school) = $\dfrac{\text{number planning to go to graduate school}}{46 \text{ total students}} = \dfrac{15}{46}$

(b) Use $P(A') = 1 - P(A) = 1 - \dfrac{15}{46} = \dfrac{31}{46}$

or P(not planning to go to graduate school) = $\dfrac{\text{number planning not to go to graduate school}}{\text{total number of students}} = \dfrac{31}{46}$

(c) P(take another math course but not planning to go to graduate school) = $\dfrac{18}{46} = \dfrac{9}{23}$

(Probability)

Grade Yourself

Circle the question numbers that you had incorrect. Then indicate the number of questions you missed. If you answered more than three questions incorrectly, you will have to focus on that topic. If a topic has fewer than three questions and you had at least one wrong, we suggest you study that topic. Read your textbook or a review book or ask your teacher for help.

Subject: Sequences and Other Topics

Topic	Question Numbers	Number Incorrect
Sequences	1, 2, 3, 4, 5, 6, 7	
Series	8, 9, 10, 11, 12, 13, 14, 15, 16	
Arithmetic sequences and series	17, 18, 19, 20, 21, 22, 23, 24, 25, 26	
Geometric sequences and series	27, 28, 29, 30, 31, 32, 33, 34, 35, 36, 37, 38, 39, 40	
Mathematical induction	41, 42, 43, 44, 45	
The Binomial Theorem	46, 47, 48, 49, 50, 51, 52, 53, 54, 55, 56, 57, 58	
Counting principles	59, 60, 61, 62, 63, 64	
Probability	65, 66, 67	